英文商務契約導讀

吳仲立　律師
Joyce C.Y.Huang
編著

國家圖書館出版品預行編目資料

英文商務契約導讀／吳仲立, Joyce C. Y. Huang編著.
－－初版四刷.－－臺北市：三民，2008
面； 公分
參考書目：面
ISBN 978-957-14-3634-0 (平裝)

1. 商業英語-應用文 2. 契約(法律)

493.6 91011753

英文商務契約導讀

編著者 吳仲立 Joyce C. Y. Huang
發行人 劉振強
發行所 三民書局股份有限公司
地址／臺北市復興北路386號
電話／(02)25006600
郵撥／0009998-5
印刷所 三民書局股份有限公司
門市部 復北店／臺北市復興北路386號
重南店／臺北市重慶南路一段61號
初版一刷 2002年8月
初版四刷 2008年8月
編 號 S 585100
行政院新聞局登記證局版臺業字第〇二〇〇號

ISBN 978-957-14-3634-0 (平裝)

編著者序

在可預期的未來，當台灣加入世界貿易組織(WTO)後，國內產業結構勢必發生極大衝擊與轉變，而WTO的機制將打破區域與國界的藩籬，企業界的生存與競爭模式亦將日新月異，為因應此潮流，再加上英文乃世界公認的通用語言之事實，台灣企業必須【英文化】，方能邁向【國際化】，進而始有【全球化】佈局的可能；如此，台灣企業在國際商業舞台上才能鞏固現有及增進未來的競爭地位。

本書期待為國內企業在國際法商事務英文化的工程上，盡一份心力；然而，我們並不以此為滿足，隨後一系列有關國際法商事務的實用叢書，也將陸續與讀者見面，至盼各界前輩及後起之秀，不吝給予我們批評與指正。

惠聯地政商務法律事務所

國際法商事務部

吳仲立律師・Joyce C. Y. Huang

西元2001年1月

英文商務契約導讀

目　次

◈第一章

解讀英文契約基本結構

壹、契約名稱／標題(Title)

一、契約名稱／標題一覽表

英文契約之種類眾多不勝枚舉，以下僅列舉部分契約名稱／標題作為參考範例。

㈠買賣契約

就商品買賣契約之名稱／標題，可有以下表現方式：

1.	Agreement	協議書；契約；同意書
2.	Confirmation of Purchase	購買確認（書）
3.	Confirmation of Sale (order)	銷售（訂購）確認（書）
4.	Export Contract	出口契約
5.	Export and Import Contract	進出口契約
6.	Import Contract	進口契約
7.	Order Sheet	訂單
8.	Purchase Agreement	購買協議書；購買契約；購買同意書
9.	Purchase Confirmation	購買確認（書）
10.	Purchase Contract	購買契約
11.	Purchase Note	購買備忘錄
12.	Purchase Order	購買訂單
13.	Sales Agreement	銷售協議書；銷售契約；銷售同意書
14.	Sales and Purchase Contract	買賣契約

15.	Sales Confirmation	銷售確認（書）
16.	Sales Contract	銷售契約
17.	Sales Note	銷售備忘錄
18.	Trade Agreement	貿易協議書；貿易契約；貿易同意書
補充小站		

㈡經銷契約

就經銷契約之名稱／標題，可有以下表現方式：

1.	Agreement	協議書；契約；同意書
2.	Distributorship Agreement	經銷協議書；經銷契約；經銷同意書
3.	Distributorship Contract	經銷契約
補充小站		

獨家／非獨家 (Exclusive / Non–Exclusive)
經銷／代理 (Distributorship / Agency)

㈢技術合作契約／專利授權契約

就技術合作契約之名稱／標題，可有以下表現方式：

1.	Agreement	協議書；契約；同意書
2.	Technical Assistance Agreement	技術援助契約；技術援助協議書；技術援助同意書
3.	Technological Assistance Agreement	技術援助契約；技術援助協議書；技術援助同意書
4.	Technical Collaboration Agreement	技術合作契約；技術合作協議書；技術合作同意書
5.	License Agreement	許可證契約；許可證協議書；許可證同意書（授權契約書）
6.	Patent License Agreement	專利許可證契約；專利許可證協議書；專利許可證同意書（專利授權契約書）
7.	Technical Information Agreement	技術情報契約；技術情報協議書；技術情報同意書
8.	Exchange of Technical Information Agreement	技術情報交換契約；技術情報交換協議書；技術情報交換同意書
9.	Patent / Royalty Agreement	專利及權利金契約
補充小站		

㈣合資契約

就合資契約而言，其名稱 / 標題，可有以下表現方式：

1.	Agreement	協議書；契約；同意書
2.	Joint Venture Agreement	合資經營契約；合資經營協議書；合資經營同意書
補充小站		

㈤租賃契約

就租賃契約而言，其名稱 / 標題，可有以下表現方式：

1.	Agreement	協議書；契約；同意書
2.	Lease	租賃契約；租約
3.	Lease Agreement	租賃協議書；租賃契約；租賃同意書
補充小站		

㈥分期付款契約

就分期付款契約而言，其名稱／標題，可有以下表現方式：

1.	Agreement	協議書；契約；同意書
2.	Installment Payment Agreement	分期付款協議書；分期付款契約；分期付款同意書
補充小站		

注意！！ 有無差異 (Installment / Instalment)

㈦附擔保之讓與契約

就附擔保之讓與契約而言，其名稱／標題，可有以下表現方式：

1.	Agreement	協議書；契約；同意書
2.	Assignment With Warranties	附擔保讓與契約
補充小站		

㈧貸款契約

就貸款契約而言，其名稱／標題，可有以下表現方式：

1.	Agreement	協議書；契約；同意書
2.	Loan Agreement	貸款協議書；貸款契約；貸款同意書
補充小站		

㈨僱傭契約

就僱傭契約而言，其名稱／標題，可有以下表現方式：

1.	Agreement	協議書；契約；同意書
2.	Employment Agreement	僱傭協議書；僱傭契約；僱傭同意書
補充小站		

㈩保證契約

就保證契約而言，其名稱／標題，可有以下表現方式：

1.	Agreement	協議書；契約；同意書
2.	Guaranty	保證契約；保證書
3.	Guaranty Agreement	保證協議書；保證契約；保證同意書
補充小站		

㈩不動產抵押契約

就不動產抵押契約而言，其名稱／標題，可有以下表現方式：

1.	Agreement	協議書；契約；同意書
2.	Real Estate Mortgage Agreement	不動產抵押契約；不動產抵押協議書；不動產抵押同意書
補充小站		

注意！！
動產／不動產 (Personalty / Realty)
動產／不動產 (Personal estate / Real estate; Real property)

㈢營業秘密契約

就營業秘密契約而言，其名稱／標題，可有以下表現方式：

1.	Agreement	協議書；契約；同意書
2.	Trade Secret Non–disclosure Agreement	營業秘密契約；營業秘密協議書；營業秘密同意書
3.	Non–discolosure and Confidentiality Agreement	保密合約；保密協議
補充小站		

㈣信託契約

就信託契約而言，其名稱／標題，可有以下表現方式：

1.	Agreement	協議書；契約；同意書
2.	Trust Agreement	信託契約；信託協議書；信託同意書
3.	Fiduciary Contract	信託契約
補充小站		

注意 委託書 (Trust deed)

二、概念區辨

㈠Contract與Agreement之區辨

大體而言，英文通常以Contract或Agreement指稱「契約」之意思。而兩者間意義之關聯性或差異性，頗令人玩味。

所謂Agreement，中文翻譯之語詞可能有：⑴契約、⑵同意、⑶協議、⑷一致、⑸符合；根據Steven H. Gifis編撰之法律字典，其將Agreement定義為：「由二或二個以上合法適格者，相互意思表示一致之表明，其通常地導致一契約之發生。」(“a manifestation of mutual assent between two or more legally competent persons which ordinarily leads to a contract”, Steven H. Gifis, Law Dictionary)。

至於Contract一詞，中文一般翻譯為「契約」(又稱「合約」或「合同」)；而Steven H. Gifis編撰之法律字典，則將Contract定義為：「契約乃指一個或一組之意思表示，一旦違反時，法律賦予救濟；或其對此意思表示之履行，法律在某些情況下視為一種義務者。」(“a promise, or set of promises, for breach of which the law gives a remedy, or the performance of which the law in some way recognized as a duty”, Steven H. Gifis, Law Dictionary)。

上述有關Agreement（合意）與Contract（契約）的差異，在

美國統一商法典 § 1–201 <11>亦表露無遺:「所謂契約，係指由當事人間之合意所衍生，經法律及其他應適用之一切法律規範而賦予法律意義之權利義務之總合。」

注意!!
Agreement的定義為:「由二或二個以上合法適格者，相互意思表示一致之表明，其通常地導致一契約之發生。」
Contract的定義為:「契約乃指一個或一組之意思表示，一旦違反時，法律賦予救濟；或其對此意思表示之履行，法律在某些情況下視為一種義務者。」

然而，Steven H. Gifis之法律字典，進一步說明Agreement之含意，略以:「在通常使用上，Agreement乃比Contract、Bargain、Promise更廣泛的名詞，Agreement包括買賣、贈與、與其他財產的移轉，與Promise相同，不負擔法律上義務。當Agreement作為Contract同義字時，部分權威人士將其限縮於相互意思表示一致。」(“In common usage, it is a broader term than contract, bargain, or promise, since it includes executed sales, gifts, and other transfers of property, as well as promises without legal obligation While agreement is often used as a synonym for contract, some authorities narrow it to mean only mutual assent.”, Steven H. Gifis, Law Dictionary)。

由上述得知， 1. Agreement的意義較Contract為廣泛，且Agreement不當然須負擔法律上的義務。 2. Agreement通常作為Contract之同義字，部分權威人士將Agreement限縮為相互意思表示一致。雖然，在概念上得明確區辨，但實務在使用上，似乎未做嚴格區分。

不過須附帶一提的是，Agreement與Contract在學理上會被認為有所差異，主要是在英美法系國家。英美法系認為要使合意(Agreement)成為有效之契約(Contract)，尚必須具備約因(consideration)、明確性原則(Doctrine of vagueness)及書面性(Formal contract or Contract by deed)等要求。有關這些概念的說明，我們將另以專書介紹。

注意！！

區辨Contract與Agreement：

1. Agreement僅為Contract構成要素之一。
2. Steven H. Gifis之法律字典，將二者定義如下：
   (1)Agreement：「事實上的協議」。
   (2)Contract：「當事人間合意所生之全部法律上義務」。
3. 嚴格說來，二者有異，然而實務未加區分。

(二)其他常見之書面合意

1. Letter of Intent（實務上有人簡稱為L / I）：得翻譯為「意願書」、「意向書」等等

Letter of Intent（意向書）只是一種概念，在於指稱某種文書的性質，僅在於表明寫作者或簽署者願意就某一種重要的交易或安排進行談判或簽約的意向而已，一項文書是否具備Letter of Intent的性質，與該項文書的名稱或標題是否寫成Letter of Intent，

無必然的關係；一項名為備忘錄（Memorandum）的文書，其性質也有可能只是一種Letter of Intent。

談到Letter of Intent（意向書）的性質，到底其性質為何呢？根據Steven H. Gifis法律字典之闡述，**「Letter of Intent（意向書）非為一種契約，並且不構成一種具有拘束力之契約。更甚者，Letter of Intent（意向書）係為當事人暫時意願之表達，並且對當事人不生任何義務。」**（"This letter is not a contract, and it does not constitute a binding agreement. Rather, it is "an expression of tentative intentions of the parties", and creates no liability as between the parties.", Steven H. Gifis, Law Dictionary）。**易言之，Letter of Intent（意向書）本來的意義並非有如契約般拘束當事人之效力，僅得視為一種針對某種交易的意向暫時的意願表達。**

不過，實務上許多名為Letter of Intent的文書，究其內容因為已有相當明確的具體約定，以致於其性質可能被判定為是一種「契約」、「預約」或「草約」，這是不得不留意的地方。所以日本人把Letter of Intent翻譯成「預備的合意」，自有其道理。

2. Memorials：得翻譯為「議事錄」

Memorials（議事錄）係指當事人於協商過程中，將其協商要旨以議事錄之方式記載。其屬於訂定正式契約前之中間產物。有疑問的是，它是否有拘束當事人的效力。**英美契約法認為，議事錄非謂當事人有「契約的意思」，遽而認定具有契約之效力。是否構成契約，不應僅就名稱之形式而認定，仍需就其內容加以綜合判斷。**

注意！！ minutes亦有議事錄之意，但通常係指正式會議所作成之文字記錄。

3. Memorandum（複數為Memoranda；Memorandums；縮寫為memo.；mem.)：得翻譯為「備忘錄」

根據Steven H. Gifis法律字典之闡述，Memorandum（備忘錄)：「係屬一種非正式紀錄；一種交易之書面摘要書或未來契據之輪廓；係一種依簡單、扼要之格式表述之契據」。("an informal record; a brief note, in writing, of some transaction or an outline of some intented instrument; an instrument drawn up in brief and compendious form.", Steven H. Gifis, Law Dictionary)。然而，當備忘錄以契約之形態作成時，則與契據 (instrument) 具有相同之意義，不得以備忘錄之字樣為藉口，而否認其契約之效力，在實務上常有爭議，不容小覷。

貳、說明 (Premises) 與前言 (Preamble)

一、締約日期 (Date of Signing)

㈠結　構

1. 主要結構一

This Agreement (is) made and entered into in 地點（可省略） *this* 日期（序數＋*day of* 月,年）.

例：本契約於2001年10月10日**當日**訂定。

This Agreement made and entered into this tenth (10th) day of October, 2001.

enter into在此有「締結」之意，例如To enter into a contract.

2. 主要結構二

This Agreement (is) made and entered into in 地點（可省略） *as of* 日期（序數＋*day of* 月,年）.

例：本契約於2001年10月10日**當時**訂定。

This Agreement made and entered into as of tenth (10th) day of October, 2001.

㈡說 明

1. **締約日期放在前文者**，即便與實際締約日期不同，法律上仍以前揭所示締約日期視為締約日期，而訂約日期原則上為契約之生效日 (Effective Date)。
2. **締約日期放在契約簽名之末端者**，除非另有規定外，應以簽名末端的日期為締約日期。
3. **當事人欲使契約溯及生效者**，則使用主要結構2：當時 (as

of) 而非使用當日 (this)。契約的生效日，以記載的日期為生效日。

4. **實務上常見締約日期之欄位空白，委由他方填入。**此種情形可能生有不利於己之危險，因此，最好能事先協議並填入締約日期，以避免他方惡意利用而造成不利己的後果。

貼心小幫手

月　份	日　序	
1月 January	1日 first (1st)	18日 eighteenth (18th)
2月 February	2日 second (2nd)	19日 nineteenth (19th)
3月 March	3日 third (3rd)	20日 twentieth (20th)
4月 April	4日 fourth (4th)	21日 twenty-first (21st)
5月 May	5日 fifth (5th)	22日 twenty-second (22nd)
6月 June	6日 sixth (6th)	23日 twenty-third (23rd)
7月 July	7日 seventh (7th)	24日 twenty-fourth (24th)
8月 August	8日 eighth (8th)	25日 twenty-fifth (25th)
9月 September	9日 ninth (9th)	26日 twenty-sixth (26th)
10月 October	10日 tenth (10th)	27日 twenty-seventh (27th)
11月 November	11日 eleventh (11th)	28日 twenty-eighth (28th)
12月 December	12日 twelfth (12th)	29日 twenty-ninth (29th)
	13日 thirteenth (13th)	30日 thirtieth (30th)
	14日 fourteenth (14th)	
	15日 fifteenth (15th)	

	16日　sixteenth (16th) 17日　seventeenth (17th)	31日　thirty-first (31st)

二、締約地點 (Place of Signing)

如未訂定準據法或訴訟管轄法院者，則締約地點的法律即為解決紛爭的重要依據。但在國際貿易實務上，契約當事人未在同一地點簽約，實屬常見。可能先由當事人之一方審閱簽名（與封印）後，再交由他方簽名（與封印）。因此，當事人簽署契約未必於同一地點下，則可省略締約地點。

三、締約當事人 (Signing Parties) 及其住所或主營業所 (Address / Principal Office of Signing Parties)

㈠結　構

1. 主要結構一：（締約當事人為雙方時）

by and between A公司（*hereinafter called* “A”） *and* B公司（*hereinafter called* “B”）.

例：由A（股份）有限公司（以下簡稱A）與XYZ（股份）有限公司（以下簡稱XYZ）雙方當事人（訂定）。

by and between A Co. Ltd.（hereinafter called A）and

XYZ Co. Ltd.（hereinafter called XYZ）.

2. 主要結構二：（締約當事人為三方時）

by and among A公司（*hereinafter called* "A"） *and* B公司（*hereinafter called* "B"）*and* C公司 （*hereinafter called* "C"）.

例：由X（股份）有限公司（以下簡稱"X"）、Y（股份）有限公司（以下簡稱"Y"）與Z（股份）有限公司（以下簡稱"Z"）三方當事人（訂定）。

by and among X Co. Ltd. (hereinafter called "X") and Y Co. Ltd. (hereinafter called "Y") and Z Co. Ltd. (hereinafter called "Z").

㈡說明

1. hereinafter called = hereinafter referred to as（以下簡稱……)

2. 我國公司法第二條第二項規定：「公司名稱應標明公司之種類」。例如：無限公司、兩合公司、有限公司、股份有限公司。在美國，公司名稱後面需加"Inc."、"Corporation"、或"Incorporation"等字樣；在英國，公司名稱後面需加"Ltd."或"Limited"等字樣，以表明公司種類。

No	英　文	中　文
1.	herein	在此；鑒於
2.	hereafter	此後
3.	hereinafter	在下文中；以下
4.	hereinbefore	在上文中；以上
5.	hereunder	在下文中；在此之下

No	英　文	中　文
1.	hereat	因此
2.	hereby	特此；由此
3.	herefrom	由此；從此
4.	herein	在此；鑒於
5.	hereof	關於此點
6.	hereon	於此
7.	hereto	至此
8.	herewith	附此；藉此

No	英　文	中　文	縮　寫
1.	company	公司；商號	Co., co.
2.	corporation	（美）公司；股份有限公司；法人團體	Corp., corp.
3.	（美）Incorporated	股份有限公司的；按股份公司組織的	Inc.
	（英）Limited	股份有限公司的；按股份公司組織的	Ltd.
	（法）Société Anonyme	股份有限公司的；按股份公司組織的	S.A.
	（西德）Gesellschaft mit beschränkter Haftung	股份有限公司的；按股份公司組織的	G.m.b.H.
	（義大利）Società a responsabilità limitata	股份有限公司的；按股份公司組織的	S.a..r.l. or S.r.l.
4.	（美）Co., Inc.	股份有限公司	
5.	（英）Co., Ltd.	股份有限公司	

No	英　文	中　文	縮　寫
1.	A sole proprietorship	（美）獨資企業	
	A partnership	（美）合夥企業	
	A corporation	（美）公司	
	Joint–stock company	（美）合夥；合股事業 （英）股份有限公司的一種	
2.	Limited company	有限公司；有限責任公司；股份有限公司	
	Limited liability company	股份有限公司；責任有限公司	
	Company limited by shares	股份有限公司	
	Private limited company	私人有限公司（類似我國之有限公司）	Pte.
3.	（英） Public limited company	公共有限公司（向公眾募股之一種有限公司）	p.l.c. plc. PLC.
	（法） Société Anonyme	公共有限公司（向公眾募股之一種有限公司）	S.A.
	（德） Aktiengesellschaft	公共有限公司(向公眾募股之一種有限公司)	A.–G.
	（荷）	公共有限公司（向公眾	N.V.

	Naamloze Vennootschap	募股之一種有限公司)	
	(義大利) Società per Azioni	公共有限公司(向公眾募股之一種有限公司)	S.p.A
4.	Unlimited company	無限公司	

四、締約當事人設立之國籍 (Nationality)

㈠結　構

1. 主要結構一

a corporation duly organized and existing under the laws of 某國家*, having its principal office at* 住址.

注意!!
正式地;正當地 (duly)
經正式授權的 (duly authorized)

2. 主要結構二

a 某國家的 *corporation, with its principal place of business at* 住址.

3. 主要結構三

a 某國家的 *corporation having its registered office at* 住址.

※以主要結構一為例：

一公司係依美國加州法而組織設立，其主營業所位於（住址）。

a corporation duly organized and existing under the laws of the State of California, U.S.A., having its principal office at 住址.

㈡說　明

1. 在單一法域國家，當事人表明其設立所依據之所在國法律之方式較無問題，但在一國數法之國家，各法域各有所屬法律，因此需明確表明該所屬法域，而不得單以該國家表明其設立所依據之法律。例如公司是依美國加州法所設立，則宜以"Under the Laws of the State of California"方式表示，而非以"Under the Laws of U.S.A."方式表示。因為在一國數法之美國，並不存在"Laws of U.S.A."。
2. **主要結構一、主要結構二、與主要結構三均可代用。**

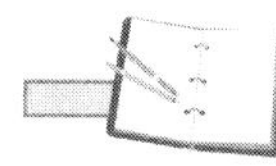

貼心小幫手

No	英　文	中　文	縮　寫
1.	Headquarters	總部；總辦事處	
	Head office	總公司；總辦事處	H.O.
	Central office	總公司	
	Principal office （注意：不要誤寫成 “principle”）	總公司；主事務所	
2.	Registered office	（公司）註冊登記之辦事（公）處	
3.	branch office	分公司；分支機構	
	a Filiale	（法國）分公司；分行	
	a daughter–company	（法國）分公司	
	a branch company	分公司	
	a subsidiary company	子公司；附屬公司	

五、締約緣由 (Recitals / Whereas Clause)

㈠主要結構

WITNESSETH
WHEREAS,..,
and
WHEREAS, ...,
NOW THEREFORE, in consideration of ....,it is agreed as follows:

例：

證明

鑒於A願指定B，

並

B願接受該指定為獨占銷售的經銷商

因此，以前述與雙方同意以誠信履約為約因，茲同意如下：

WITNESSETH

WHEREAS, A desires to appoint B,

and

WHEREAS,B desires to accept such appointment, as an exclusive sales agent.

NOW THEREFORE, in consideration of the premises and the mutual conventions to be faithfully performed herein contained, it is agreed as follows:

㈡說　明

1. WITNESSETH係指witness（「證明」）之古體字，為第三人稱單數，時態為現在式。其主詞為"This Agreement"。
2. WHRERAS clause（緣由條款），即"WHEREAS..........as follows："，係屬"WITNESSETH"之受詞。

<table>
<tr><td>

THIS AGREEMENT,.....................................

（主詞）

WITNESSETH

（動詞－第三人稱單數、現在式）

WHEREAS,...,

and

WHEREAS, ...,

（受詞）

結構說明：

主詞 *(THIS AGREEMENT)*
＋動詞 *(WITNESSETH)*
＋受詞 *(WHEREAS..., and WHEREAS....)*

</td></tr>
</table>

3. in consideration of（以……為約因）。根據Steven H. Gifis法律字典說明，「約因係為契約之動機。約因為契約成立要件之一，生有拘束當事人之效力；如欠缺時，則不得強制當事人履行。」的意旨。(“the inducement to a contract, something of value given in return for a performance or a promise of performance by another, for the purpose of forming a contract; one element of a contract that is generally required to make a promise binding and to make the agreement of the parties enforceable as a contract. To find consideration there must be a performance or a return promise which has been bargained for by the parties.”, Steven H. Gifis, Law Dictionary)

六、前言範例

契約

本契約由A公司（以下簡稱A）——總公司位於住址，依照某國之法律設立；與B公司（以下簡稱B）——總公司位於住址，依照某國之法律設立，於＿年＿月＿日，地點，訂定契約。

茲証明

鑒於＿＿＿＿＿＿＿＿＿＿

與

鑒於 ______________________________

因此，以 ____________ 為約因，同意如下：

AGREEMENT

THIS AGREEMENT, (is) made and entered into in 地點 this 日期（序數）day of 月, 年 by and between A (hereinafter called “A”), a corporation duly organized and existing under the laws of 某國家, having its principal office at 住址, and B (hereinafter called “B”), a corporation duly organized and existing under the laws of 某國家, having its principal office at 住址.

WITNESSETH

WHEREAS,..,

and

WHEREAS, ...,

NOW THEREFORE, in consideration of ...., it is agreed as follows:

參、本文 (Body / Operative Part)

一、定義規定 (Definition)——用語定義

定義條款顧名思義，其作用乃在於定義契約書所出現之特定用語，以釐清該用語於契約之範圍與內容，其攸關當事人之權利義務甚鉅。當事人於審酌契約時，應多加留意，否則可能導致自身權利的限縮與排除。例如：「服務項目」(“Service”)之範圍，一經限定後，該公司未來可能受同種業務服務競業禁止的限制；又如「技術情報」(Technical Information) 所涵蓋範圍一經定義後，得命當事人就保密範圍為一定之作為或不作為。以上看似簡單的定義性條款，卻潛藏對締約當事人業務活動範圍的限制。契約中的每個小環節，都能左右當事人權利義務得、喪、變更。因此，每字每句，都不能等閒視之。

就定義條款之引言文字，舉例說明如下：

> 定　義
>
> 除本契約內容有明定之其他意義者外，下列名詞有如下之定義：
>
> DEFINITIONS
>
> The following terms shall have the following meanings unless otherwise clearly required by the context:

二、實質規定
——契約實質內容 (Basic Conditions)

實質規定為契約的重頭戲，並堪稱契約最精華部分。**其主要規定當事人於該契約類型的權利義務關係。**不同契約種類，其所訂定之實質內容也將有極大的差異。在總論書中，無法將各契約的實質規定一一臚列。但在附錄二，我們有擬定實質規定的例示條款，供讀者參考。(參見pp.147～149)

三、一般規定 (General Terms and Conditions)

一般規定原則上不因契約種類的不同，而有顯著的差異。易言之，一般規定幾乎得適用於各種契約，具有共通性。常見之一般規定，大致如下：

㈠**契約有效期間與終止** (Duration / Period / Term, and Termination)

契約有效期間從契約生效日 (Effective Day) 到契約屆滿日 (Expiry Date) 或終止日 (the day of Termination) 為止。大致而言，契約上所載之締約日期或當事人簽署契約之日期，為契約之生效日。締約當事人自契約生效日起，就契約上所載之文義，行使權利、負擔義務。為明確規定契約之有效期間，得以下列文字表述：

期　間
本契約之有效期間，自契約生效日起一年。
Duration
The term of this agreement shall be for a period of one (1) year, beginning on the Effective Date.

而契約因期間之屆滿或終止，失其效力，係屬當然。唯值得注意者，契約提前終止之事由可能有以下兩種情形，一為法定終止事由；一為意定終止事由。而法定終止事由，可能因各國法律之不同，而有不同規範內容；至於意定終止事由則得由當事人訂定，然不得違反法律。

注意!! 有些契約並不因某段期間之經過或契約義務履行之完成而失效，例如房屋買賣契約，買賣雙方各履行交付價金及移轉標的物之義務後，契約即處於「睡眠」狀態，而非失效，否則一旦幾年後發現房屋有賣方故意不告知之瑕疵，買方如何依據一個失效之契約主張權利呢?

再者，契約關係因期間之屆滿而消滅，但有時為續約之便，通常會加註優先續約權或相關續約的規定，以展延繼續往來的商業合作關係。展延（續約）的相關文字敘述，例如:

例一：

續　約

於本契約屆滿之六十日前，當事人得以誠信更新雙方均得接受之本契約之條件。

Renewal

For the period commencing sixty (60) days prior to the termination of the initial term of this Agreement, the parties may negotiate in good faith to renew this Agreement on terms acceptable to both parties.

例二：

續約

在契約有效期間後，除另有較早終止本契約之情形者外，本契約自動延長一年。

Renewal

This agreement will automatically renew for successive one (1) year of terms, unless terminated sooner as provided by the terms of this agreement.

㈡不可抗力（法語：Force Majeure；英語：Act of God）

我國民法第二百二十五條第一項規定：「因不可歸責於債務人之事由，致給付不能者，債務人免給付義務」。該條規定係法律為

衡平因情事變更所致之債務不履行。於當事人非出於故意，且非締約當時所得預料之情形下，免除當事人依債之本旨履行契約之義務。

由於「不可歸責於債務人之事由」係屬較廣泛、籠統之概念，依照各國法律與實務之解釋下，其所包括之範圍與法律效果也不盡相同。為避免無謂的紛爭，應明確訂定該範圍、法律效果與其發生時之通知等。而「不可歸責於債務人之事由」應包含「不可抗力」之概念。因此，「不可抗力」條款常見於一般條款中，免除當事人給付之義務。就「不可抗力」條款，舉例說明如下：

不可抗力

因天災、戰爭、內亂、暴動、國內的緊急危難、罷工、封港、暴風雨、地震、其他自然力、或不可歸責於任何一方當事人之事由、或因不能獲得原料及運輸所致之債務不履行或遲延，受影響之當事人得於該事故發生完結後之一定合理期限內補正，而得免除於進一步之履行，但該一定期間不得超過六十日。當有上述情事發生時，當事人A應以商業上可適用的慣例，立即以書面通知當事人B，但該期間不得超過二十個營業日，並且出具發生該不可抗力事故的證明文件。

Act of God (Force Majeure)

In the event of any failures or delays by either party hereto in the performance of all or any part of this agreement due to acts of

God, war, riot, insurrection, national emergency, strike, embargo, storm, earthquake, or other natural forces, or by the acts of anyone not a party to this agreement, or by the inability to secure materials or transportation, then the party so affected shall be executed from any further performance for a period of time after the occurrence as may reasonably be necessary to remedy the effects of that occurrence, but in no event more than sixty (60) days. If any of the stated events should occur, Party A shall promptly notify Party B in writing as soon as commercially practicable, but in no event more than twenty (20) business days and provide documentation evidencing such occurrence.

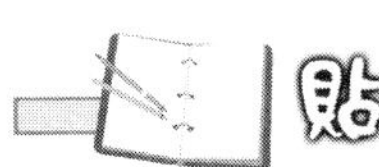

貼心小幫手

不可抗力事故之種類範例

編號	種類	中文說明
1.	lockouts	工廠封鎖
2.	acts of God	天災
3.	war	戰爭
4.	fire	火災
5.	riot	內亂
6.	civil commotion	民眾騷擾
7.	insurrection	暴動

8.	national emergency	國內的緊急危難
9.	strike	罷工
10.	labor trouble or other industrial disturbances	勞動爭議或其他僱傭之糾紛
11.	embargo	禁運
12.	storm	暴風雨
13.	tempest	颶風
14.	flood	洪水
15.	drought	旱災
16.	earthquake	地震
17.	other natural forces	其他自然力
18.	any other events beyond the control of the parties	不可歸責於任何一方當事人之事由
19.	blockade	封港
20.	governmental interference	政府干涉
21.	legal restriction	法律限制
22.	plague, pestilence, epidemic	瘟疫；流行病；傳染病
補充小站		

㈢轉讓 (Assignment)

契約上所定之權利義務是否得轉讓於第三人，我們試就債權讓與、債務承擔與契約轉讓，分成三個部分說明。

1. 債權讓與

這是純粹就基於契約所獲得之權利轉讓他人而言。根據我國民法第二百九十四條規定：

「債權人得將債權讓與於第三人。但左列債權，不在此限：

一、依債權之性質，不得讓與者。

二、依當事人之特約，不得讓與者。

三、債權禁止扣押者。」

再者，雖然得將債權讓與於第三人，但依我國民法第二百九十七條第一項之規定，債權讓與人或受讓人仍應踐行通知債務人之程序，否則對於債務人不生效力。（民法第二百九十七條：「債權之讓與，非經讓與人或受讓人通知債務人，對於債務人不生效力。但法律另有規定者，不在此限。受讓人將讓與人所立之讓與字據提示於債務人者，與通知有同一之效力。」）

2. 債務承擔

這是純粹將基於契約所產生之義務轉由他人承擔之謂。有關債務承擔之訂定，得參照我國民法第三百條與第三百零一條所規定之方式為之。

民法第三百條：「第三人與債權人訂立契約承擔債務人之債務者，其債務於契約成立時，移轉於該第三人。」

民法第三百零一條：「第三人與債務人訂立契約承擔其債務者，非經債權人承認，對於債權人，不生效力。」

由上述法條得知，第三人得與債權人或債務人訂定債務承擔契約。但需注意者，第三人與債務人所訂定之債務承擔契約，應

得債權人之同意。

3. **契約轉讓**

這是將整個契約之權利及義務，包括契約當事人之地位轉讓予他人之謂。我國民法並無直接適當的條文可供參照；不過基於契約自由之原則，只要契約當事人不反對，且其轉讓不違反有關法令，契約轉讓亦無不可。

由於契約上所定之權利義務是否得轉讓於第三人，倘若未明定許可、禁止或限制，極易引起紛爭之發生，故最好於契約中明定具體轉讓之條款。就轉讓條款，得以下列文字表述：

例一：

轉　讓

當事人之一方未得他方書面同意前，不得將本契約或與本契約有關之權利義務讓與第三人。**除有正當理由者外，他方不得拒絕同意。**

Assignment

Neither party may assign this agreement or any right or obligation under this agreement to any third party, without prior written consent of the other party. **The other party shall not withhold such consent without justifiable reason.**

黑體字係當自己處於「應得他方當事人同意」的劣勢下，仍得限縮他方當事人同意的權限。易言之，他方

當事人不得單憑其一己之好惡，任意拒絕同意轉讓契約，而使我方遭受不利的結果。

例二：

轉　讓

當事人之一方未得他方明示書面之同意前，不得轉讓契約之全部或一部。該同意係屬他方當事人之單獨絕對自由判斷。任何違反本條之轉讓，應視為無效。

Assignment

Neither party may assign this agreement in whole or in part without the express written permission of the other party, which consent shall be in the sole and absolute discretion of the party. Any assignment made in contravention of this provision shall be deemed void.

例三：

轉　讓

未得B書面同意前，A不得轉讓本契約與本契約之權利；B得轉讓或移轉其權利與義務於其繼承人或受讓人（不論係因股票或資產之購買、公司合併、或其他），但是，當B向A直接競爭者轉讓契約時，A應有權利終止契約而不負擔義務。

Assignment

This Agreement and the rights hereunder are not assignable by A without the prior written consent of B; B may assign or transfer its rights and obligations to any successor or assignee (whether by purchase of stock or assets, merger, or otherwise), provided however in the event of an assignment of this Agreement by B to a direct competitor of A, A shall have the right to terminate the Agreement without liability.

㈣修改 (Amendment / Modification)

當事人於締約過程中，難免隨著情事變更或營業政策改變，而更動契約的內容。而契約一經雙方當事人同意並簽訂後，對雙方即形成拘束力。

但鑒於發生情事變更或營業政策改變的情形，並不限於締約前，於締約後，仍有存在的可能。一般而言，當事人於契約中，將訂定修改條款 (Modification Clause)。當事人得依照修改條款 (Modification Clause) 所定的方式，以書面（即修改書）補充原契約的不足或瑕疵。該修改書的內容，得由雙方另議訂定。以下亦就修改條款，舉例說明：

修　改

僅得經由雙方當事人正式簽署之書面，而修改本契約。

Modification

This Agreement may be modified only in a writing duly signed by each party.

㈤完整合意條款 (Entire Agreement)

正式締約前，當事人可能曾有預備締約之口頭合意，甚至書面合意，如備忘錄 (memorandum)、議事錄 (memorials) 等。為確保正式契約成為當事人行使權利的唯一依據，應表明契約條款完全取代之前所有之預備之口頭及書面合意。舉例說明如下：

完整合意

本契約之條文乃當事人完整的合意內容，以取代當事人締約前有關本契約內容（不論口頭或書面）之提議、協商、對談與討論之一部或全部。

Entire Agreement

This Agreement constitutes the entire agreement between the parties pertaining to the subject matter hereof, and fully supersedes any and all prior proposals (oral or written), negotiations, conversations, and discussions between or among the parties relating to the subject matter of this Agreement.

如契約書中訂有完整合意之條款時，似乎該契約書係為唯一履約之依據。但締約後，若當事人堅持其他書面合意內容之效力優於原契約之部分條款時。唯一補救方式，則可於該書面合意之

內容加註「本書面合意之條款優先於……之效力」。例如：

本備忘錄之規定，優於雙方當事人於____年____月____日所簽訂契約之第____條（第____項、第____款）。 The provisions in this Memorandum take precedence to Article…of the agreement between the parties dated （日期）.

㈥翻譯／語言 (Translation / Language)

締約當事人分屬不同國籍與不同語系時，通常其所使用的語言文字，也隨之不同。締約當事人之一方如收受他方當事人所提供的契約，而該契約所使用之文字與其本國文字不同時，當事人可能採取以下兩種方式，以解決因語言文字不同，以致無法真正瞭解契約內容或因此造成對契約認知差距的問題。

其一、將他國語言的契約版本，翻譯為本國語言的契約版本。但譯本僅供參考用，雙方仍僅簽署該外文契約。如簽約時，僅以外文契約為簽訂對象，則僅該外文契約有拘束雙方當事人的效力。例如：

語　文 當事人雙方以英文簽署本契約，並且雙方當事人不受其他契約譯本的拘束。 Language This agreement has been executed by the parties hereto in the

English Language, and no translated version of this agreement into other languages shall be controlling or binding upon any of parties hereto.

其二、將外文契約翻譯為本國語言契約譯本，而譯本非僅供參考，當事人簽約以兩種語言版本同時簽訂。但由於同時以不同語言版本簽約，如外文契約與譯本間發生歧異時，究以何種版本為主，則具爭議性。因此，為迅速解決因契約所生的爭執，及避免該窘境的發生，當事人應明訂翻譯條款或語言條款，允許或禁止譯本的使用。如允許譯本的使用，當版本間有歧異時，應載明以何種版本的效力為優先。

例如：

翻　譯

如提供契約譯本予甲方時，甲方同意依照所有英文契約之條款與條件，並且版本間如有衝突時，以英文版本為主。

Translation

In the event a translated version of this Agreement is provided to Party A, Party A understands and agrees that this English version, and all terms and conditions found herein, controls and that any conflict between the two shall be resolved in favor of this English version.

㈦存續 (Survival)

所謂存續條款，係規定締約當事人之一方，於契約屆滿或終止後，就某些契約條款，仍繼續享有不受侵害的一定利益。這些條款例如：競業禁止、保密義務、保證等相關規定。惟應注意者，存續條款之期間、內容與範圍應合理、明確，不宜過度濫用而限制他方於市場公平競爭的機會。存續條款之表達如下：

> 例一：
>
> 存　續
>
> 本契約中之第__條之規定，於本契約終止後，仍有效力。
>
> Survival
>
> The obligations under Article__will survive any termination of this agreement.

> 例二：
>
> 存　續
>
> 儘管本契約期間或續約期間終止或屆滿，茲同意這些權利義務仍屬存續，包括但不限於前述，與以下第__條之規定。
>
> Survivability
>
> Notwithstanding the termination or expiration of the term of this Agreement or any renewal period thereof, it is acknowledged and agreed that those rights and obligations shall survive, including, without limiting the foregoing, the following provisions in Article__.

㈧分離 (Severability)

分離條款係指契約部分條文無效，該無效之條文並不影響其他條文的有效性。該條文的訂定，得防止惡意當事人規避契約責任。

例如：

> 分　離
>
> 法院判決如認本契約任何條款之一部或全部無效者，不影響其他有效之部分條款之執行力。
>
> Severability
>
> A judicial determination that any provision of this agreement is invalid, in whole or in part, shall not affect the enforceability of those provisions unaffected by the finding of inability.

㈨不棄權 (Non–Waiver)

當事人之一方基於維持繼續性的商業合作關係，可能在他方當事人第一次債務不履行的情形下，不立即終止契約，而選擇宥恕 (即放棄依據契約得享有的權利)。如他方當事人再次發生相同之違約情事，根據此條款之訂定下，不因前所為之宥恕 (棄權)，而喪失依據契約所得主張的權利。該條款之說明如下：

> 例一：
>
> 不棄權
>
> 當事人之一方於任何時期，就本契約條款發生債務不履行

或債務遲延者，並不當然解釋為此方當事人放棄執行該條款的權利。

Non–Waiver

The failure or delay of any party to enforce at any time any of the provisions hereof will not be construed to be a waiver of the right of such party thereafter to enforce any such provision.

例二：

不棄權

當事人之一方於任何時期，無法請求他方當事人履行本契約任何規定者，並不影響此後於任何時期，其請求他方履行之所有權利。當事人之一方放棄對他方違約請求之權利，不視為接受、解釋為或保留對該條款本身權利之放棄、或對此後任何違約請求之放棄、或對其他條款權利之放棄。

Non–Waiver

The failure of either party at any time to require performance by the other party of any provision hereof shall not affect in any way the full rights to require such performance at any time thereafter. The waiver by either party of breach of any provisions hereof shall not be taken, construed, or held to be a waiver of the provision itself or a waiver of any breach thereafter or a waiver of any other provision hereof.

(十)通知 (Notice)

根據我國民法之規定，某些情形下，當事人所為之法律行為，自始、當然、確定的無效。該無效法律行為固毋庸通知。而某些情形下，當事人所為之法律行為，則屬效力未定。例如限制行為能力人未得法定代理人之允許，所訂立之契約，需經法定代理人之承認，始生效力（我國民法第七十九條）；表意人意思表示出於錯誤、或不自由（例如被詐欺、脅迫），原則上，於一定期間內，得撤銷之（我國民法第八十八條、第九十二條）。而該撤銷與承認之方式，我國民法第一百十六條規定，應以意思表示為之。意思表示似乎不限於書面，口頭亦無不可。但是以口頭通知他方當事人，容易產生是否完成通知的疑問。在法庭上，自己應就已通知對造的事實，負舉證責任，而自己如能適時地提供書面通知的證據，將不致因該舉證責任，而受敗訴的判決結果。

其他法律上規定之催告（我國民法第二百五十四條）、解除契約（我國民法第二百五十八條）、終止契約（我國民法第二百六十三條）等方式，亦均屬最好以書面進行較有利於舉證的情形。

書面通知的方式種類眾多，如郵寄、掛號、電報、電傳、傳真、與發電子郵件等等。且該通知是否一經發出即生通知之效力（採發信主義），抑或於他方當事人收受時，方生通知的效力（採到達主義），不無疑問。因此，就通知的方式應予確認，並且就通知的生效時點亦應予載明。以下就通知之方式與生效時點，舉例說明：

例一：

通　知

本契約之通知將以親自交付、藉由公認係良好傳輸之傳真、或藉由掛號信件郵寄，送達於本契約簽名處上所載之當事人地址或根據本契約第__條通知修改後之當事人地址。藉由郵寄所為之通知，於該郵件交寄後五個日曆天，視為已送達。

Notice

Notice under this agreement will be sufficient if hand delivered, delivered by facsimile with acknowledgment of good transmission, or if mailed by certified or registered mail, to the parties at the addresses first set forth on the signature page of this agreement or as amended by notice pursuant to Article__. Notice by mail will be deemed received five (5) calendar days after deposit.

例二：

通　知

一切通知、請求、詢問或根據本契約之其他通訊方式，應以書面為之。當以親自交付，或以美國掛號信件郵寄後5日，或交寄於全國公認，夜間無休之特別信差後2日，且附有簽名之收據，並向以下雙方當事人之住址投遞時，應視為已正式即時送達：

甲方　　　　　　　　　　乙方

地址　　　　　　　　　　地址

____________________　　____________________

Notice

All notices, requests, demands and other communications hereunder shall be in writing and shall be deemed to have been duly given immediately upon personal delivery or five (5) days after mailing by U.S. certified or registered mail; or two (2) days after deposit with any nationally recognized overnight courier, with written verification of receipt, and addressed to the parties at the address set forth below:

To Party A　　　　　　　　To Party B

ADDRESS　　　　　　　　ADDRESS

____________________　　____________________

㈩仲裁 (Arbitration)

當事人以訴訟解決紛爭，係為常見的方式。但鑒於訴訟往往更增加當事人在勞力、時間、與費用方面的負擔之考量，再加上法官就涉及專門性知識的案件，未必具備充分專業的審查能力。因此，當事人可能採取較為迅速、經濟、保密性與專業性兼具的仲裁方式，以解決紛爭。而一般仲裁條款的內容，不外乎表明同意交付仲裁字樣、仲裁範圍、仲裁地點、仲裁機構、仲裁所適用

的法規、與仲裁費用的負擔等等。

例如：

仲　裁

由本契約所引起或與本契約有關之糾紛、爭議或歧異或違約情事，應依紐約州仲裁法與美國仲裁協會仲裁規則，在紐約市仲裁，仲裁人所為之仲裁判斷有終局判斷之效力，雙方當事人應受其拘束，仲裁費用由敗方（受不利仲裁判斷之當事人）負擔。

Arbitration

Any dispute, controversy or difference arising out of or relating to this contract or the breach thereof, shall be settled by arbitration in the City of New York, in accordance with the arbitration law of the State of New York and under the rules of the American Arbitration Association, whose award shall be final and binding on both parties and whose expenses shall be borne by the parties against whom the award is made.

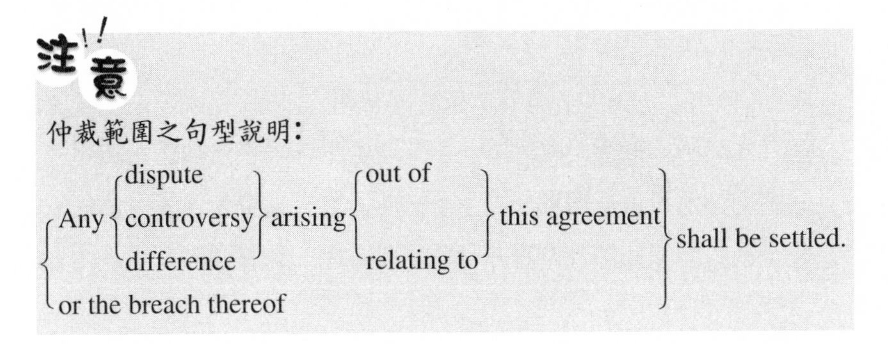

㈢準據法 (Applicable Law / Governing Law) 與訴訟管轄 (Jurisdiction)

當事人得選擇上述仲裁的方式外，亦得選擇追求慎重而正確的訴訟方式，解決因契約所產生的紛爭。訴訟所採的準據法與訴訟管轄原則上應為同一。惟應注意者，法院判決後，對於當事人於法院所在地的財產是否得以執行，應為當事人考量選用何國準據法與何國訴訟管轄的首要因素。以下就準據法與訴訟管轄選定的表達方式，舉例說明：

例一：

準據法與訴訟管轄

本契約之解釋、履行應按照中華民國法律。雙方當事人茲同意以台灣台北地方法院為管轄法院。

Governing Law and Jurisdiction

This agreement shall be construed and enforced in accordance with the laws of the Republic of China. The parties hereby consent to and submit to the jurisdiction of Taiwan Taipei District Court.

例二：

準據法與訴訟管轄

本契約之效力、履行與一切有關本契約之事務均遵照加州法為準據法並依此解釋。如生爭議時，雙方當事人同意以位於舊金山之加州北部管區聯邦地方法院為唯一管轄法院。

Governing Law and Jurisdiction

The validity, performance, and all other matters pertaining to this agreement will be governed and construed in accordance with the laws of the State of California. In the event of any dispute, the parties submit to the exclusive jurisdiction in the United States District Court for the Northern District of California in Francisco.

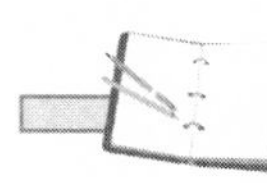

貼心小幫手

美國聯邦地方法院一覽表

州名	縮寫	地方法院英文名稱	地方法院中文名稱
Alabama (阿拉巴馬)	M.D. Ala.	U.S.District Court for the Middle District of Alabama	阿拉巴馬中區聯邦地方法院
	N.D. Ala.	U.S.District Court for the Northern District of Alabama	阿拉巴馬北區聯邦地方法院
	S.D. Ala.	U.S.District Court for the Southern District of Alabama	阿拉巴馬南區聯邦地方法院
Alaska (阿拉斯加)	D. Alas	U.S.District Court for the District of Alaska	阿拉斯加地區聯邦地方法院

Arizona (亞利桑那)	D. Ariz.	U.S.District Court for the District of Arizona	亞利桑那地區聯邦地方法院
Arkansas (阿肯色)	E.&W.D. Ark	U.S.District Court for the Eastern and Western Districts of Arkansas	阿肯色東部暨西區聯邦地方法院
California (加利福尼亞)	C.D. Cal	U.S.District Court for the Central District of California	加利福尼亞中央區聯邦地方法院
	E.D. Cal	U.S.District Court for the Eastern District of California	加利福尼亞東區聯邦地方法院
	N.D. Cal	U.S.District Court for the Northern District of California	加利福尼亞北區聯邦地方法院
	S.D. Cal	U.S.District Court for the Southern District of California	加利福尼亞南區聯邦地方法院
Colorado (科羅拉多)	D. Colo.	U.S.District Court for the District of Colorado	科羅拉多地區聯邦地方法院

Connecticut（康乃狄克）	D. Conn.	U.S.District Court for the District of Connecticut	康乃狄克地區聯邦地方法院
Delaware（德拉瓦）	D. Del.	U.S.District Court for the District of Delaware	德拉瓦地區聯邦地方法院
District of Columbia（哥倫比亞特區）	D.D.C.	U.S.District Court for the District of Columbia	哥倫比亞特區地區聯邦地方法院
Florida（佛羅里達）	M.D. Fla.	U.S.District Court for the Middle District of Florida	佛羅里達中區聯邦地方法院
	N.D. Fla.	U.S.District Court for the Northern District of Florida	佛羅里達北區聯邦地方法院
	S.D. Fla.	U.S.District Court for the Southern District of Florida	佛羅里達南區聯邦地方法院
Georgia（喬治亞）	M.D. Ga.	U.S.District Court for the Middle District of Georgia	喬治亞中區聯邦地方法院
	N.D. Ga.	U.S.District Court for the Northern District of Georgia	喬治亞北區聯邦地方法院
	S.D. Ga.	U.S.District Court for the Southern	喬治亞南區聯邦地方法院

		District of Florida	
Guam （關島）	D. Guam	U.S.District Court for the District of Guam	關島地區聯邦地方法院
Hawaii （夏威夷）	D. Hawaii	U.S.District Court for the District of Hawaii	夏威夷地區聯邦地方法院
Idaho （愛達荷）	D. Idaho	U.S.District Court for the District of Idaho	愛達荷地區聯邦地方法院
Illinois （伊利諾）	E.D. Ill.	U.S.District Court for the Eastern District of Illinois	伊利諾東區聯邦地方法院
	N.D. Ill.	U.S.District Court for the Northern District of Illinois	伊利諾北區聯邦地方法院
	S.D. Ill.	U.S.District Court for the Southern District of Illinois	伊利諾南區聯邦地方法院
Indiana （印第安那）	N.D. Ind.	U.S.District Court for the Northern District of Indiana	印第安那北區聯邦地方法院
	S.D. Ind.	U.S.District Court for the Southern District of Indiana	印第安那南區聯邦地方法院
Iowa （愛荷華）	N.D. Iowa	U.S.District Court for the Northern	愛荷華北區聯邦地方法院

		District of Iowa	
	S.D. Iowa	U.S.District Court for the Southern District of Iowa	愛荷華南區聯邦地方法院
Kansas （堪薩斯）	D. Kan.	U.S.District Court for the District of Kansas	堪薩斯地區聯邦地方法院
Kentucky （肯塔基）	E.D. Ky.	U.S.District Court for the Eastern District of Kentucky	肯塔基東區聯邦地方法院
	W.D. Ky.	U.S.District Court for the Western District of Kentucky	肯塔基西區聯邦地方法院
Louisiana （路易斯安那）	E.D. La.	U.S.District Court for the Eastern District of Louisiana	路易斯安那東區聯邦地方法院
	M.D. La.	U.S.District Court for the Middle District of Louisiana	路易斯安那中區聯邦地方法院
	W.D. La.	U.S.District Court for the Western District of Louisiana	路易斯安那西區聯邦地方法院

Maine (緬因)	D. Me.	U.S.District Court for the District of Maine	緬因地區聯邦地方法院
Maryland (馬里蘭)	D. Md.	U.S.District Court for the District of Maryland	馬里蘭地區聯邦地方法院
Massachusetts (麻薩諸塞)	D. Mass.	U.S.District Court for the District of Massachusetts	麻薩諸塞地區聯邦地方法院
Michigan (密西根)	E.D. Mich.	U.S.District Court for the Eastern District of Michigan	密西根東區聯邦地方法院
	W.D. Mich.	U.S.District Court for the Western District of Michigan	密西根西區聯邦地方法院
Minnesota (明尼蘇達)	D. Minn.	U.S.District Court for the District of Minnesota	明尼蘇達地區聯邦地方法院
Mississippi (密西西比)	N.D. Miss.	U.S.District Court for the Northern District of Mississippi	密西西比北區聯邦地方法院
	S.D. Miss.	U.S.District Court for the Southern District of	密西西比南區聯邦地方法院

		Mississippi	
Missouri (密蘇里)	E.D. Mo.	U.S.District Court for the Eastern District of Missouri	密蘇里東區聯邦地方法院
	W.D. Mo.	U.S.District Court for the Western District of Missouri	密蘇里西區聯邦地方法院
Montana (蒙大拿)	D. Mont.	U.S.District Court for the District of Montana	蒙大拿地區聯邦地方法院
Nebraska (內布拉斯加)	D. Neb.	U.S.District Court for the District of Nebraska	內布拉斯加地區聯邦地方法院
Nevada (內華達)	D. Nev.	U.S.District Court for the District of Nevada	內華達地區聯邦地方法院
New Hampshire (新罕布夏)	D.N.H.	U.S.District Court for the District of New Hampshire	新罕布夏地區聯邦地方法院
New Jersey (新澤西)	D.N.J.	U.S.District Court for the District of New Jersey	新澤西地區聯邦地方法院
New Mexico (新墨西哥)	D.N.M.	U.S.District Court for the District of New Mexico	新墨西哥地區聯邦地方法院
New York (紐約)	E.D.N.Y.	U.S.District Court for the Eastern	紐約東區聯邦地方法院

		District of New York	
	N.D.N.Y.	U.S.District Court for the Northern District of New York	紐約北區聯邦地方法院
	S.D.N.Y.	U.S.District Court for the Southern District of New York	紐約南區聯邦地方法院
	W.D.N.Y.	U.S.District Court for the Western District of New York	紐約西區聯邦地方法院
North Carolina (北卡羅來納)	E.D.N.C.	U.S.District Court for the Eastern District of North Carolina	北卡羅來納東區聯邦地方法院
	M.D.N.C.	U.S.District Court for the Middle District of North Carolina	北卡羅來納中區聯邦地方法院
	W.D.N.C.	U.S.District Court for the Western District of North Carolina	北卡羅來納西區聯邦地方法院
North Dakota	D.N.D.	U.S.District Court	北達科他地區聯

（北達科他）		for the District of North Dakota	邦地方法院
Ohio（俄亥俄）	N.D.O.	U.S.District Court for the Northern District of Ohio	俄亥俄北區聯邦地方法院
	S.D.O.	U.S.District Court for the Southern District of Ohio	俄亥俄南區聯邦地方法院
Oklahoma（奧克拉荷馬）	E.D. Okl.	U.S.District Court for the Eastern District of Oklahoma	奧克拉荷馬東區聯邦地方法院
	N.D. Okl.	U.S.District Court for the Northern District of Oklahoma	奧克拉荷馬北區聯邦地方法院
	W.D. Okl.	U.S.District Court for the Western District of Oklahoma	奧克拉荷馬西區聯邦地方法院
Oregon（奧瑞岡）	D. Or.	U.S.District Court for the District of Oregon	奧瑞岡地區聯邦地方法院
Pennsylvania（賓夕法尼亞）	E.D. Pa.	U.S.District Court for the Eastern District of Pennsylvania	賓夕法尼亞東區聯邦地方法院

	M.D. Pa.	U.S.District Court for the Middle District of Pennsylvania	賓夕法尼亞中區聯邦地方法院
	W.D. Pa.	U.S.District Court for the Western District of Pennsylvania	賓夕法尼亞西區聯邦地方法院
Puerto Rico (波多黎各)	D.P.R.	U.S.District Court for the District of Puerto Rico	波多黎各地區聯邦地方法院
Rhode Island (羅德島)	D.R.I.	U.S.District Court for the District of Rhode Island	羅德島地區聯邦地方法院
South Carolina (南卡羅來納)	D.S.C.	U.S.District Court for the District of South Carolina	南卡羅來納地區聯邦地方法院
South Dakota (南達科他)	D.S.D.	U.S.District Court for the District of South Dakota	南達科他地區聯邦地方法院
Tennessee (田納西)	E.D. Tenn.	U.S.District Court for the Eastern District of Tennessee	田納西東區聯邦地方法院
	M.D. Tenn.	U.S.District Court for the Middle District of	田納西中區聯邦地方法院

		Tennessee	
	W.D. Tenn.	U.S.District Court for the Western District of Tennessee	田納西西區聯邦地方法院
Texas（德克薩斯）	E.D. Tex.	U.S.District Court for the Eastern District of Texas	德克薩斯東區聯邦地方法院
	N.D. Tex.	U.S.District Court for the Northern District of Texas	德克薩斯北區聯邦地方法院
	S.D. Tex.	U.S.District Court for the Southern District of Texas	德克薩斯南區聯邦地方法院
	W.D. Tex.	U.S.District Court for the Western District of Texas	德克薩斯西區聯邦地方法院
Utah（猶他）	D. Ut.	U.S.District Court for the District of Utah	猶他地區聯邦地方法院
Vermont（佛蒙特）	D.V.	U.S.District Court for the District of Vermont	佛蒙特地區聯邦地方法院
Virgin Islands（維爾京群島）	D.V.I.	U.S.District Court for the District of Virgin Islands	維爾京群島地區聯邦地方法院
Virginia	E.D. Va.	U.S.District Court	維吉尼亞東區聯

(維吉尼亞)		for the Eastern District of Virginia	邦地方法院
	W.D. Va.	U.S.District Court for the Western District of Virginia	維吉尼亞西區聯邦地方法院
Washington (華盛頓)	E.D. Wash.	U.S.District Court for the Eastern District of Washington	華盛頓東區聯邦地方法院
	W.D. Wash.	U.S.District Court for the Western District of Washington	華盛頓西區聯邦地方法院
West Virginia (西維吉尼亞)	N.D.W. Va.	U.S.District Court for the Northern District of West Virginia	西維吉尼亞北區聯邦地方法院
	S.D.W. Va.	U.S.District Court for the Southern District of West Virginia	西維吉尼亞南區聯邦地方法院
Wisconsin (威斯康辛)	E.D. Wis.	U.S.District Court for the Eastern District of Wisconsin	威斯康辛東區聯邦地方法院
	W.D. Wis.	U.S.District Court for the Western	威斯康辛西區聯邦地方法院

		District of Wisconsin	
Wyoming (懷俄明)	D. Wyo.	U.S.District Court for the District of Wyoming	懷俄明地區聯邦地方法院

肆、後文──結語

一、結尾文句(Witness clause)

就結尾條款舉例說明如下:

> 本契約一式二份，由雙方正式授權其承辦人員或代表，
>
> 於 { 首開日期 / (該日期) } 訂定生效。
>
> IN WITNESS WHEREOF, this agreement has been executed by the duly authorized officers or representatives of the parties in duplicate as of the date { first above written. / this day of ______(日期)______. }

二、簽名 (Signature)

以簡單明瞭之方式編排簽名的位置，為現代契約的共同特性。

因此，以下就法人代表人簽約之編排方式，舉簡單之範例：

甲方：×××公司	乙方：×××公司
簽名	簽名
姓名：×××	姓名：×××
職稱	職稱
日期	日期
封 印	封 印
Party A:	Party B:
Signature	Signature
Printed Name	Printed Name
Title	Title
Date	Date
L.S. / Seal	L.S. / Seal

注意！！ L.S.＝*Locus Sigilli*＝place of the seal

伍、契約格式一覽表

契　約

本契約由A公司（以下簡稱A）──總公司位於住址，依照某國之法律設立；與B公司（以下簡稱B）──總公司位於住址，依照某國之法律設立，於＿＿年＿＿月＿＿日，地點，訂定契約。

茲証明

鑒於＿＿＿＿＿＿＿＿＿＿

與

鑒於＿＿＿＿＿＿＿＿＿＿

因此，以＿＿＿為約因，同意如下：

第一條

第二條

第三條

⋮

本契約一式二份，由雙方正式授權其承辦人員或代表，於首開日期訂定生效。

甲方：×××公司	乙方：×××公司
簽名	簽名
姓名：×××	姓名：×××
職稱	職稱
日期	日期
封印	封印

AGREEMENT

THIS AGREEMENT, (is) made and entered into in 地點 this 日期（序數）day of 月, 年 by and between A (hereinafter called "A"), a corporation duly organized and existing under the laws of 某國家, having its principal office at 住址, and B (hereinafter called "B"), a corporation duly organized and existing under the laws of 某國家, having its principal office at 住址.

WITNESSETH

WHEREAS,..,

and

WHEREAS, ...,

NOW THEREFORE, in consideration of ...., it is agreed as follows:

Article 1

⋮

IN WITNESS WHEREOF, this agreement has been executed by the duly authorized officers or representatives of the parties in duplicate as of the date first above written.

Party A: ________	Party B: ________
________	________
Signature	Signature
________	________
Printed Name	Printed Name
________	________
Title	Title
________	________
Date	Date
L.S. / Seal	L.S. / Seal

陸、英文契約基本條款
——中英對照檢查表 (Check List)

編號	檢查項目	是	否	不適用	備註
一	契約名稱／標題 (Title)				
二	締約日期 (Date of signing)				
三	締約地點 (Place of signing)				
四	締約當事人 (Signing parties) 及其住所或主營業所 (Address / Principal Office of signing parties)				
五	締約當事人設立之國籍 (Nationality)				
六	締約緣由 (Recitals / Whereas clause)				
七	本文定義條款 (Definition)				
八	本文實質規定 (Basic Conditions)				
九	本文一般規定 (General terms and conditions)				
	⑴契約有效期間與終止 (Duration / Period / Term, and Termination)				
	⑵不可抗力 (Force Majeure;				

	Act of God)				
	(3)轉讓 (Assignment)				
	(4)修改 (Amendment / Modification)				
	(5)完整合意條款 (Entire Agreement)				
	(6)翻譯／語言 (Translation / Language)				
	(7)存續 (Survival)				
	(8)分離 (Severability)				
	(9)不棄權 (Non–Wavier)				
	(10)通知 (Notice)				
	(11)仲裁 (Arbitration)				
	(12)準據法與訴訟管轄 (Applicable Law / Governing Law and Jurisdiction)				
十	結語、簽名 (Witness clause / Signature)				

◈第二章

撰寫英文契約的要領

一、積極提昇法商英文能力

只要您不是離群索居，或自外於這個競爭激烈的世界，那麼您將驚覺到現在已是一個法律與商務結合的時代。我們每天的生活無可避免地依循著現代法商的遊戲規則在運轉著。法律與商業的影響力日日在左右我們的思考模式，來不及因應的，只有漸被時代的洪流淹沒。

在我國加入WTO之此時此刻，提昇國人的法商英文能力已是刻不容緩；而在國際商務交流將更趨頻繁的可預知未來，國人接觸英文契約及其他英文法商文件的機會亦勢必愈多。因此，積極提昇自己的法商英文能力，厥是撰寫一份好的英文契約所應具備的首要能力。

二、利用現有同類型的契約範例

利用現有的或可供參考的範例，能節省撰寫契約者的勞力、時間及費用。雖然相類似或同種類契約有助於撰寫契約者更容易完成契約書的擬定，但若未仔細推敲並比較該條文對自己的適合度及優劣性，則更可能導致不利於己的情形發生。例如：一般條款中有關翻譯條款（或語言條款）內容，有時條款內容規定，雙方當事人得同時簽訂中英版本並存的契約，但發生歧異時，以其中一種語言的契約為準；有時條款內容規定，一契約版本排除其他譯本的效力，即該契約之譯本僅供參考（詳見第41頁－（六）翻譯／語言）。到底哪一種規定方式較切合己方的需要且合乎己方

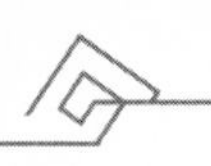

的利益，如未仔細推敲、斟酌及擇用，將可能使自己招致不利。

值得注意的是，選用相似或同種類契約範例作為擬定契約書之參考時，首先需釐清該契約範本先天上傾向之立場為何，俾就利己且合理之事項納入契約書中。再者，各契約範本所使用之文字敘述方式各有所長，故應選用通順易懂的文體作為表述，方可避免誤解契約內容的發生。

三、力求契約架構健全

為了使契約的條文便於閱讀、理解及執行上的管理，在撰寫契約時，應加強契約架構的健全。撰寫契約時，契約名稱應適當；條文的排列應分明且合乎邏輯；每個條文的標題亦應簡短有力，俾收畫龍點睛之效。當然，一份好的契約一定要言之有物，段落分明，最好一項權利或義務即規定成一項或一款，不要過度冗長。

四、充分表達當事人意圖

契約製作的好壞，常取決於該契約是否能充分、清晰與妥適的表達當事人意圖。契約如能完全表達當事人意圖，當事人於契約上的爭執，可相對的減少。因此，契約糾紛的起因之一，往往在於當事人未能以正確充分的法律文字，表達其意圖所致。

由於科技的快速發展，交易型態的複雜及多樣，一份契約要能充分表達當事人的真意，除了要有精鍊的文字表達能力外，許多科技、工商的術語及行話，亦應完全瞭解，始能正確地運用在合約中；當然，足夠的實務（商場）經驗，更有助於賦予一份契約的說服力!

五、條款用語前後一致性

契約擬定者有時一味追求契約的詳細與完整，而盡可能將有利於己的條款納入契約書中，但卻忽略條文彼此間之規範意旨，是否生有矛盾與衝突，乃至於定義條款所定義的文字與其他條款所出現要表達同一意義的文字不是使用同一個單字或名詞，而發生混淆的情形。例如，貨品有時以“goods”稱之、或以“materials”、“merchandise”稱之。雖然，在實質上，這三個文字所蘊含的意義並不完全相同，但若是所表達的意義為同一時，最好使用同一個文字或名詞，避免混淆。

六、考量未來可能的變動因素

當事人締約後，在契約履行過程中，可能因情事的變更，是當事人所始料未及的，倘若仍強使當事人依原契約履行義務，反而有失公平。這種可能發生情事變更的情形，在國際商務契約益形明顯。當事人得於擬定契約條款時，特別注意讓部分條款具有相當的彈性，使當事人在一定情形下，亦不致因突遭情事的變更而讓自己受到不公平的待遇。

七、定義性條款應清晰完整

定義性條款目的在於避免契約書中同一事物的反覆說明與爭議。一個完整、清晰的定義條款，將有助於一份契約複雜內容的

簡化與清楚；定義條款所描述的內涵，應與契約其他條款所出現指涉相同意義的文字相呼應，其重要性已如第五點所言。且定義文字或詞句與日後當事人權利義務的履行息息相關。例如：「技術情報」(technologic information) 一詞在契約內所定義的內涵，牽涉到當事人於保密條款中所應保密的範圍；亦與當事人依照義務條款中所應提供協力的範圍有相當大的關連。

八、最後應再次檢查所擬定之條款

契約條款大致完成後，應就下列事項再作檢查及確認：

1. 契約之當事人是否適格？正確？
2. 條款內容是否仍有缺漏？
3. 條款順序排列上是否妥適？
4. 條款間是否互相牴觸？
5. 定義條款是否完整？並且是否與其他條款所出現指涉相同意義的文字呼應？
6. 條款內容是否違反我國公序良俗與法律強制禁止的規定？

針對上述事項檢查確認後，最後仍宜請教法律暨商務的專業工作者，對於已擬定之契約條款，再次通盤檢查、審認，以確保契約的妥當性與完整性。

相信不少人因忽略契約的重要性而吃虧。而現代企業經營者不僅應專注於業績與利潤的創造，更應著重於損失風險的規避。相信正在研讀此類書籍的人都存有減低風險的觀念。恭喜聰明的你，已向成功邁進一大步了!!

◈第三章

英文契約常用字、詞彙與練習

一、推敲慣用語

(一)above-mentioned	*adj.* 前述的；上述的
=aforementioned	
=aforesaid	
=foregoing	

<例句>

鑒於前述之緣由與雙方之同意，當事人同意如下：

NOW, THEREFORE, in consideration of the above-mentioned, and of the mutual consents and agreements, the parties agree as follows:

(二)hereby	*adv.* 因此；據此；由此
【說明】“here＋介系詞” (here解釋為「本契約」)	
=介系詞 this agreement	

<例句>

甲方基此承認且同意本產品係遵照本契約規定而製造。

Party A hereby acknowledges and agrees that Products manufactured pursuant to this agreement.

㈢herein　　*adv.* 在此處

【說明】"here＋介系詞"（here解釋為「本契約」）

＝介系詞 this agreement

<例句>

甲方 (Party A) 想要基於本契約之規定與條件，向乙方 (Party B) 購買本契約所述之商品。

Party A desires to purchase from Party B products as described herein, under the terms and conditions of this agreement.

㈣hereinafter　　*adv.* 以下；在下文中

【說明】"here＋in＋after"解釋為「以後在本契約中」。

<例句>

本文件，包括所有的附件、附錄，以下係共同屬於本契約的參考。

This document, including all Exhibits and Appendices attached, are hereinafter collectively referred to as the agreement.

appendices為appendix的複數形

㈤hereof	*adv.* 於此；關於此點
【說明】“here＋介系詞”（here解釋為「本契約」）＝介系詞 this agreement	

＜例句＞

當事人之一方於任何時期，無法請求他方當事人履行本契約任何規定者，並不影響此後於任何時期，其請求他方履行之所有權利。

The failure of either party at any time to require performance by the other party of any provision hereof shall not affect in any way the full rights to require such performance at any time thereafter.

㈥hereto	*adv.* 在此；關於這個
【說明】“here＋介系詞”（here解釋為「本契約」）＝介系詞 this agreement	

＜例句＞

語文：本件當事人係以英文簽署本契約，而翻譯成其他語文之契約版本並不生拘束本件雙方當事人之效力。

Language：This agreement has been executed by the parties hereto in the English Language, and no translated version of this agreement into other languages shall be controlling or binding upon any of parties hereto.

(七)hereunder　　　*adv.* 在此之下；在下文；依此

（＝under this）

【說明】"here＋介系詞"（here解釋為「本契約」）

＝介系詞 this agreement

＜例句＞

甲方同意根據本契約所授與的權利係屬非獨家的權利。

Party A acknowledges and agrees that the right are granted hereunder are nonexclusive.

(八)thereafter　　　*adv.* 其後；以後

＜例句＞

此後，本契約除另有提前終止之規定者外，本契約將自動更新續約二年期間。

Thereafter, this agreement will automatically renew for successive two (2) years of terms, unless terminated sooner as provided by the terms of this agreement.

(九)thereof　　　*adv.* 關於……；從此

＜例句＞

儘管本契約期間或續約期間終止或屆滿，茲同意這些權利義務仍屬存續，包括但不限於前述，與以下第______條之規定。

Notwithstanding the termination or expiration of the term of this

Agreement or any renewal period thereof, it is acknowledged and agreed that those rights and obligations shall survive, including, without limiting the foregoing, the following provisions in Article ______.

(十) whatsoever＝whatever	*adj.* 無論如何的

＜例句＞

無論如何任何由甲方所引起或與甲方之履行或違約有關的一切損失、損害或索賠，甲方均有義務根據上述第十條補償乙方。

Whatsoever any loss, damage, or claim arise out of or relate to Party A's performance or breach provided in this agreement, Party A shall have an obligation to indemnify Party B pursuant to Article 10 above.

二、數目

(一)以上、以下（俱連本數計算）

1. 以上（句型）

- 數目＋and / or ＋
  - above / over
  - up / upward / upwards
  - more
- not less than＋數目

2. 以下（句型）

數目＋and / or ＋
- under / below
- down
- less

㈡未滿、超過（不連本數計算）

1. any amount over and above＋數目
2. any excess over＋數目
3. a sum exceeding＋數目
4. a sum above＋數目

㈢到……為止；從……到……為止（俱連本數計算）

1. up to數目
2. from數目 to and including數目

三、期日、期間

㈠表「特定期日」：on

＜例句＞

「產品」將於2001年12月25日**當日**上市。

The Product will be placed on the market **on** December 25, 2001.

㈡表「期間始日」

1. after（始日不算入；即自翌日起算）
2. from（始日不算入；即自翌日起算）
3. commencing with（始日算入，即自該日起算）

＜例句＞

1. 當事人一方之實質上違約未於書面通知**後**的30日內（**始日不算入，即自通知之翌日起算**）補正其瑕疵時，他方得因該違約而終止契約。

   Either party may terminate this Agreement for breach if any

material breach of this Agreement by the other is not cured within thirty (30) days after written notice thereof.

2. 當事人一方之實質上違約，**自書面通知起**之30日內（**始日不算入，即自通知之翌日起算**）未為補正其瑕疵時，他方得因該違約而終止契約。

Either party may terminate this Agreement for breach if any material breach of this Agreement by the other is not cured within thirty (30) days from written notice thereof.

3. 當事人一方之實質上違約，**自書面通知起**之30日內（**始日算入，即自通知之日起算**）未為補正其瑕疵時，他方得因該違約而終止契約。

Either party may terminate this Agreement for breach if any material breach of this Agreement by the other is not cured within thirty (30) days commencing with written notice thereof.

注意！！

包括始日：
- as from
- on and after
- on or after
- on and from
- from and including

不包括始日：from and after

㈢表「期間末日」

1. before（末日不算入）
2. by（末日算入）

3. on or before（末日算入）

＜例句＞

1. 於2001年12月25日以前（末日不算入），甲方得基於單獨決定下修改本契約之條款。

   Party A may have its sole discretion to modify the terms of this agreement before December 25, 2001.

2. 於2001年12月25日以前（末日算入），甲方得基於單獨決定下修改本契約之條款。

   Party A may have its sole discretion to modify the terms of this agreement by December 25, 2001.

3. 於2001年12月25日以前（末日算入），甲方得基於單獨決定下修改本契約之條款。

   Party A may have its sole discretion to modify the terms of this agreement on or before December 25, 2001.

㈣表「期間始日及末日」

1. commencing（＝beginning）with ______ and ending with ______, both days inclusive.（該兩日均包括在內）

2. from ________ to ________, both days inclusive.（該兩日均包括在內）

3. commence on ________ and will continue until ________, both days inclusive.（該兩日均包括在內）

＜例句＞

1. 本契約期間開始於2001年12月25日，於2003年12月25日屆滿

（該兩日均包括在內）。

This agreement is in the period commencing（=beginning) with December 25, 2001, and ending with December 25, 2003, both days inclusive.

2. 本契約期間自2001年12月25日到2003年12月25日（該兩日均包括在內）。

The terms of this agreement is from December 25, 2001 to December 25, 2003, both day inclusive.

3. 本契約期間將開始於2001年12月25日，並於2003年12月25日屆滿（該兩日均包括在內）。

This agreement will commence on December 25, 2001, and will continue until December 25, 2003, both days inclusive.

四、但書規定 (proviso)

(一)but： 但是

<例句>

（中）甲方保留其商標之全部權利，但於契約期間內，授與乙方非獨家、不得移轉的權利，使用商標；但其他情形而使用甲方商標者，應事前得甲方之書面同意。

（英）Party A reserves all of its rights in its Trademarks but grants to Party B the non–exclusive, non–transferable right during the term of this Agreement to use the Trademarks, provided that any use of Party A Trademarks shall be subject to the

prior written approval of Party A.

㈡provided that ＝except that： 但是；除……之外

＜例句＞

（中）甲方保留其商標之全部權利，但於契約期間內，授與乙方非獨家、不得移轉的權利，使用商標；但其他情形而使用甲方商標者，應事前得甲方之書面同意。

（英）Party A reserves all of its rights in its Trademarks but grants to Party B the non–exclusive, non–transferable right during the term of this Agreement to use the Trademarks, **provided that** (＝**except that**) any use of Party A Trademarks shall be subject to the prior written approval of Party A.

㈢however： 然而；但是

＜例句＞

（中）關於產品促銷，甲方將與乙方協調行銷資料，**然而**，乙方有權檢查所有該資料與商標的任何使用。

（英）Party A will coordinate marketing materials with Party B relating to the promotion of the Product, **however**, Party B has the right to review all such materials and any use of the Trademarks.

㈣notwithstanding： 儘管

＜例句＞

（中）**儘管**本契約期間或續約期間的終止或屆滿，玆同意這些權利義務仍為存續，包括但不限於前述，與以下第______條之規定。

（英）Notwithstanding the termination or expiration of the term of this Agreement or any renewal period thereof, it is acknowledged and agreed that those rights and obligations shall survive, including, without limiting the foregoing, the following provisions in Article ______ .

五、代用字、同義字

㈠all / any / each　全部；任何；每個

<例句>

（中）各當事人得經向他方發書面通知，毋需理由地終止本契約。

（英）Each party may terminate this agreement without cause, by giving written notice to the other party.

㈡alter / amend / change / modify　變更

<例句>

（中）當事人不得更改、修改、或變更本契約，除非該更改、修改、或變更係經雙方當事人以書面並簽署的方式為之。

（英）This agreement shall not be changed, modified or amended by the parties of this agreement provided that such change, modification or amendment shall be in writing and be signed by both parties.

㈢agree / covenant　**同意**

＜例句＞

（中）除非雙方當事人另有書面**同意**者外，（否則）買方向賣方所為之一切付款均以美金為之。

（英）Unless otherwise agreed upon in writing between the parties, all payments from Buyer to Seller will be made in U.S. dollars.

㈣as / when　**當**

＜例句＞

（中）**當**到期款項有未付款的情形時，甲方每月應按照未付差額之1%繳付利息，但法律有較高利率之規定者，從其規定。

（英）In the event payments are not paid when due, Party A shall pay interest on the unpaid balance at the lower of one percent (1%) per month or the highest rate permitted by law.

㈤assign / convey / transfer　**移轉**

＜例句＞

（中）若未事先經他方當事人之同意，當事人任何一方均不得**轉讓**並／或**移轉**本契約之全部或一部予任何的個人、商行或公司。

（英）Neither party shall assign and / or transfer this agreement in whole or in part to any individual, firm, or corporation without the prior consent of the other party.

㈥bind / obligate　**拘束；負擔義務**

＜例句＞

（中）本件當事人係以英文簽署本契約，而翻譯成其他語言之契約版本並不生**拘束**本件雙方當事人之效力。

（英）This agreement has been executed by the parties hereto in the English Language, and no translated version of this agreement into other languages shall be controlling or bind ing upon any of the parties hereto.

㈦conclusive / final　**終局；最終**

＜例句＞

（中）該仲裁判斷應為**最終**並對雙方當事人有拘束力。

（英）The award made upon such arbitration shall be final and binding upon both parties.

㈧consider / deem　**認為；視為**

＜例句＞

（中）每批裝運應**視為**個別且獨立的契約。

（英）Each shipment shall be deem ed as a separate and independent agreement.

㈨complete / full　**完全**

＜例句＞

（中）付款方式為預付，全部契約金額，均按下列方式之一支付之：

（甲）電匯或信匯。

（乙）銀行匯票。

（丙）支票。（除非支票金額已兌現，否則不視為已經付款）

（英）Payment in advance, for full agreement amount by any of the following means:

(a) Telegraphic Transfer (T / T) or Mail Transfer. (M / T)

(b) Banker's Draft.

(c) Check. (Payment shall not be deemed received unless the amount of the check has been collected)

㈩costs / expenses　**費用**

<例句>

（中）所有與仲裁有關之**費用**應由受不利仲裁判斷之當事人負擔。

（英）All costs and expenses in relation to arbitration shall be borne by the parties against whom the award is made.

費用由……負擔 (at one's own expense)

㈯discharge / relieve　**清償；解除**

<例句>

（中）買方不得基於本契約（對賣方）提出索賠。除非該索賠係以書面業於貨物船舶到達**卸貨**港後7日內對賣方提出者外。

（英） No claim shall be made by the buyer under this agreement unless the claim is delivered in writing to the Seller within seven (7) days after the arrival of the goods carrying vessel at the port of discharge.

㈩displace / supersede　**取代**

＜例句＞

（中） 本契約應**取代**所有口頭或書面之提案，所有當事人間就有關本契約主題所為之協商、對談或討論以及所有過去之交易或商業習慣。

（英） This agreement supersedes all proposals, oral or written, all negotiations, conversations, or discussion between or among the parties relating to the subject matter of this agreement and all past dealing or industry customs.

㈫due / owing　**由於；因**

＜例句＞

（中） 買方所有付款應以美金給付，並以電匯、信匯或支票給付予賣方或其指定之人。

（英） All payments due to the Buyer shall be paid in U.S. Dollars and shall be paid by T/T, M/T, or check payable to the Seller or its designees.

㈬effective / valid　**有效**

＜例句＞

（中）本契約於生效日，由當事人正式授權的承辦人或代表人簽署。

（英）This agreement has been executed by the duly–authorized officers or representative of the parties as of the Effective Day.

㈤have access to　**有權使用**

＜例句＞

（中）當事人皆有權使用機密情報。

（英）The parties may have access to information that is confidential.

㈥exclusive / sole　**排除的；獨占的**

＜例句＞

（中）本契約有效期間內，賣方委派本契約的「經銷商」為其獨家的經銷商，而「經銷商」亦同意接受並擔任此項委派。

（英）During the effective period of this agreement, Seller hereby appoints Distributor as its exclusive distributor and Distributor accepts and assumes such appointment.

㈦furnish / supply　**提供；供應**

＜例句＞

（中）甲方將提供乙方每月之報告，內容包括訂購數量、已收到的詢價、產品最近需求預測及消費者剖析圖。

（英）Party A will furnish Party B monthly reports providing the number of orders booked, inquires received, the Product prospects currently pursued, and customer profile.

(十八)null / of no effect / void / invalid **無效的**

＜例句＞

（中）判決本契約任何條款之全部或一部**無效**者，應不影響其他有效條文之效力。

（英）A judicial determination that any provision of this agreement is invalid, in whole or in part, shall not affect the enforceability of those provisions unaffected.

(十九)mentioned / referred to **提及**

＜例句＞

（中）根據本契約，所有得或應通知的方式，應以掛號的航空郵件或電報向以下所**提及**的住址，或當事人書面通知的地址。

（英）All notices which may or shall be given under this agreement shall be made by registered airmail or cable to the address mentioned below or to such address as are notified in writing by the parties hereto.

(二十)out of **出於**

＝in connection with **關於**

＝in relation to **有關於**

＜例句＞

（中）由本契約所引起或與本契約有關的爭執、爭議、或違約，應於中華民國以仲裁方式解決之。

（英）Any dispute or difference arising <u>out of</u> or <u>relating to</u> this agreement or the breach thereof, shall be settled by arbitration in R.O.C.

(二十一)subject to / under　**根據**

＜例句＞

（中）本契約金額20%應以現金預付，並且另外80%**根據**不可撤銷信用狀所開見票即付匯票（及其匯票）支付。

（英）Twenty percent (20%) of the value of this agreement shall be paid in advance by cash and eighty percent (80%) by sight draft drawn <u>under</u> an irrevocable letter of credit.

(二十二)terms / conditions　**條件**

＜例句＞

（中）買方保留權利，以督核賣方符合本契約所訂的**條件**。

（英）Buyer reserves the right to audit Seller's compliance of the <u>terms and conditions</u> of this agreement.

(二十三)during the terms of / within　**在……期間內**

＜例句＞

（中）甲方應以自己的費用，於本契約**期間內**，提供產品正常運作的例行維修。

（英）Party A shall, at its own costs and expenses, provide routine maintenance for the proper operation of the Product, <u>during the terms of</u> this agreement.

㈤disputes / controversies / differences **爭論；爭議；歧異**

＜例句＞

（中）如中文版本與英文版本發生**爭論、爭議、歧異**時，均應以英文版本為優先。

（英）In case of disputes, controversies, differences between the Chinese version and the English version, the English shall always prevail.

㈥confidentiality/secrecy **保密；機密**

＜例句＞

（中）除本契約另有規定者外，當事人同意就技術情報**保密**，但上述（義務）不適用於：⑴______⑵______⑶______。

（英）Unless otherwise stated in this agreement, the parties agree to keep in confidentiality in technical information, provided that the preceding shall not apply to : ⑴______⑵______⑶______.

六、常見片語、生字、句子

㈠apply to **適用**

＜例句＞

（中）產品價格調漲不應**適用**於通知前，賣方已接受的訂單，亦不適用於在通知期間內，已安排裝船（運）的訂單。

（英）The price increase of the Product shall not apply to orders

accepted by Seller prior to such notice, and orders scheduled for shipment within the notice period.

㈡comply with　**遵從**

<例句>

（中）甲方同意**遵從**所有出口限制，包括但不限於美國政府所屬的相關出口限制，以及在台灣任何代理機構的國際貿易限制。

（英）Party A agrees to comply with all export restrictions, including, but not limited to, those of the government of the U.S.A. and any foreign trade restrictions by any agency in Taiwan.

㈢be entitled to　**享有……（資格或權利）**

<例句>

（中）當賣方提供更優惠的價格與條件結合方案予其他賣方的顧客時，買方應**享有**相同的折扣，該折扣溯及於賣方給與該顧客優惠價格與條件相同之日。

（英）In the event that Seller provides a combination of prices and terms which are more favorable price to another of its customers, Buyer shall be entitled to a reduction retroactive to the date that the more favorable price and terms were made available to other customers.

㈣be obligated to　**有義務**

<例句>

(中) 甲方**有義務**根據本契約提供全部技術的援助。

(英) Party A is obligated to provide full technical support according to this agreement.

㈤be responsible for **對……負有責任的**

＝be liable for

＜例句＞

(中) 甲方應為次經銷商到期的任何款項獨自**負責**。

(英) Party A shall be solely responsible and liable for any payments due to Sub–Distributor.

㈥be subject to **依照**

＜例句＞

(中) 買方保留取消或修改任何訂單的權利，而不負擔費用與義務，該取消或修改訂單將**依照**價格調漲而定。

(英) Buyer retains the right to cancel or modify any order that will be subject to a price increase, without cost or obligation.

㈦from time to time **時常**

＜例句＞

(中) 代理商應**時常**嚴格遵守任何及所有賣方對代理商的指示，並不得為任何代表、保證、承諾、締約或為其他拘束賣方的任何行為。

(英) Agent shall strictly conform to any and all instructions given by Seller to Agent from time to time and shall not make any representation, warranty, promise, contract, and agreement or

do any other act binding Seller.

㈧in accordance with　**根據；依照**

<例句>

（中）本契約條款應**依照**中華民國法律及/或法規解釋。

（英）The provisions of this agreement shall be construed <u>in accordance with</u> the laws and / or regulations of the R.O.C.

㈨including without limitation　**包括……但不限於此**

＝including but not limited to

<例句>

（中）敗訴當事人應給付勝訴當事人於訴訟或仲裁上所生之費用，**包括但不限於**法院費用（裁判費）及合理律師費。

（英）The non-prevailing party shall pay the prevailing party's costs and expense in such litigation or arbitration, <u>including, without limitation</u>, court costs, and reasonable attorneys' fees.

㈩in connection with　**與……相關**

<例句>

（中）只要基於乙方的認可之下，**有關於**產品的經銷事宜，甲方得利用次經銷商為之。

（英）Party A may use sub-distributors <u>in connection with</u> the distribution of the Product, subject to the approval of Party B.

㈩in compliance with　遵從；符合

<例句>

（中）當檢查結果，如有任何的產品或其任何零件未**符合**品質標準，則製造商應免費提供不合格品的替換予買方。

（英）If any products or any part of the products is not in compliance with the standards of quality as the result of the inspection, manufacturer shall supply buyer free of charge replacement for all parts not complying with the standards of quality.

㈪in duplicate　一式二份

in triplicate　一式三份

<例句>

（中）現行契約以**一式二份**擬定中文與英文，該兩種語言版本均有同等效力。當譯本有爭論、爭議、或歧異時，英文版本應有優先的效力。

（英）The present agreement is drawing in duplicate in the Chinese and English languages, all two versions being equally authentic. In case of any disputes, controversies, and differences of interpretation, the English version shall prevail.

㈫in its sole discretion　基於其完全的判斷（決定）

<例句>

（中）甲方得**基於其自由決斷**下，同意或不同意次經銷商。

（英）Party A may approve or disapprove of Sub–Distributors in its sole discretion.

㈣in the event of　**在……的情形；萬一；如果**

<例句>

（中）**當**甲方根據第十條終止契約**的情形時**，甲方於其自由決定下，將有權選擇履行一部或全部訂購單或取消一部或全部訂購單。

（英）In the event of termination by Party A under Article 10, Party A, in its sole discretion, will have the option to fulfill any or all purchase orders or to cancel any or all purchase orders.

㈤pursuant to　**遵照；依照**

<例句>

（中）賣方持續保證**根據**本契約所提供的產品價格，對任何顧客而言，均係為最低價格。

（英）Seller warrants at all time that the Product prices offered pursuant to this agreement are the lowest prices available to any customer.

㈥prior to　**……前**

＝before

<例句>

（中）於貨物裝船之**前**，應先空運送來兩套裝運樣品。

（英）Two sets of shipping samples to be sent by airmail prior to shipment of the goods.

㈦set forth　陳述；闡明；宣佈

＜例句＞

（中）甲方根據本契約所陳述的條件，有權直接行銷與提供產品予顧客。

（英）Party A has the right to market and offer the Product to customers directly subject to the terms and conditions set forth in this agreement.

㈧除書（除另有……者外）

1. Unless otherwise agreed upon in writing between the parties,...
   除雙方另有書面同意者外，……
2. Unless otherwise mutually agreed to by the parties,...
   除雙方另有同意者外，……
3. Unless otherwise specifies,...
   除另有特別規定者外，……

◈附錄一

各種英文契約範例條款中英對照檢查表
(Check List)

一、【買賣契約】條款檢查表

編號	檢查項目	是	否	不適用	備註
一	契約名稱／標題 (Title)				參見：p.2
二	締約日期 (Date of signing)				參見：pp.14～16
三	締約地點 (Place of signing)				參見：p.17
四	締約當事人 (Signing parties) 及其住所或主營業所 (Address / Principal Office of signing parties)				參見：pp.17～18
五	締約當事人設立之國籍 (Nationality)				參見：pp.22～23
六	締約緣由 (Recitals / Whereas clause)				參見：pp.25～26
七	本文定義條款 (Definition)				參見：p.29
八	本文實質規定 (Basic Conditions)				參見：p.30
	㈠買賣標的 (Description of goods)				
	㈡品質 (Quality)				
	㈢數量 (Quantity)				
	㈣價格 (Price)				
	㈤付款 (Payment)				
	㈥裝運 (Shipment)				
	㈦包裝 (Packing)				

	(八)箱記 (Shipping Mark)				
	(九)保險 (Insurance)				
	(十)貨物檢驗及受領 (Inspection and Acceptance of goods)				
	(土)運費保費幣值等的變動 (Fluctuations of Freight, Insurance Premium, Currency, etc.)				
	(主)稅捐 (Taxes and Duties, etc.)				
	(圭)索賠／救濟 (Claims and Remedies)				
	補充：				
九	本文一般規定 (General terms and conditions)				
	(一)契約有效期間與終止 (Duration / Period / Terms, and Termination)				參見： pp.30～31
	(二)不可抗力 (Force Majeure; Act of God)				參見： pp.32～33
	(三)轉讓 (Assignment)				參見： pp.35～38
	(四)修改 (Amendment / Modification)				參見：p.39

	㈤完整合意條款 (Entire Agreement)				參見：pp.40～41
	㈥翻譯／語言 (Translation / Language)				參見：pp.41～42
	㈦存續 (Survival)				參見：p.43
	㈧分離 (Severability)				參見：p.44
	㈨不棄權 (Non–Wavier)				參見：pp.44～45
	㈩通知 (Notice)				參見：pp.46～48
	⑪仲裁 (Arbitration)				參見：pp.48～49
	⑫準據法 (Applicable Law / Governing Law)與訴訟管轄 (Jurisdiction)				參見：pp.50～51
	補充：				
十	結語、簽名 (Witness clause / Signature)				參見：pp.63～65

二、【經銷契約】條款檢查表

編號	檢查項目	是	否	不適用	備註
一	契約名稱／標題 (Title)				獨家／非獨家 (Exclusive /Non-Exclusive) 經銷／代理 (Distributorship / Agency) 參見：p.3
二	締約日期 (Date of signing)				參見：pp.14～16
三	締約地點 (Place of signing)				參見：p.17
四	締約當事人 (Signing parties) 及其住所或主營業所 (Address / Principal Office of signing parties)				參見：pp.17～18
五	締約當事人設立之國籍 (Nationality)				參見：pp.22～23
六	締約緣由 (Recitals / Whereas clause)				參見：pp.25～26
七	本文定義條款 (Definition)				參見：p.29
	㈠經銷區域 (Territory)				

	㈡經銷商品 (Products)				
	㈢經銷權限 (Distributorship Right)				
	補充：				
八	本文實質規定 (Basic Conditions)				參見：p.30
	㈠授權 (Grant of Rights)				
	㈡經銷商承諾 (Acceptance of provisions by Distributor)				
	㈢禁止越區轉售 (Prohibition of resale outside territory)				
	㈣貨物購買價格 (Purchase price of goods)				
	㈤最低購買數量 (Minimum Purchase)				
	㈥付款 (Payment)				
	㈦交貨 (Delivery of Products)				
	㈧公司保留之權利 (Reserved Rights by Company)				
	㈨商情報告 (Information and Reports)				
	㈩維持庫存與提供售後服務				

	(Stock and after-sale service maintenance)				
	(十一)商標使用權 (Trademark Rights)				
	(十二)促銷與廣告 (Sale Promotion and Advertisement)				
	(十三)選任分經銷商 (Appointment of Sub-distributor)				
	(十四)貨物品質保證 (Warranty of Products Quality)				
	(十五)履約保證金 (Security for Performance)				
	(十六)價格變動 (Price Change)				
	(十七)保密 (Confidentiality)				
	(十八)稅捐 (Taxes and Duties)				
	補充：				
九	本文一般規定 (General terms and conditions)				
	(一)契約有效期間與終止 (Duration / Period / Terms, and Termination)				參見：pp.30～31

	(二)不可抗力 (Force Majeure; Act of God)				參見：pp.32～33
	(三)轉讓 (Assignment)				參見：pp.35～38
	(四)修改 (Amendment / Modification)				參見：p.39
	(五)完整合意條款 (Entire Agreement)				參見：pp.40～41
	(六)翻譯／語言 (Translation / Language)				參見：pp.41～42
	(七)存續 (Survival)				參見：p.43
	(八)分離 (Severability)				參見：p.44
	(九)不棄權 (Non-Wavier)				參見：pp.44～45
	(十)通知 (Notice)				參見：pp.46～48
	(十一)仲裁 (Arbitration)				參見：pp.48～49
	(十二)準據法 (Applicable Law / Governing Law)與訴訟管轄 (Jurisdiction)				參見：pp.50～51
	補充：				
十	結語、簽名 (Witness clause / Signature)				參見：pp.63～65

三、【技術合作契約】條款檢查表

編號	檢查項目	是	否	不適用	備註
一	契約名稱 / 標題 (Title)				參見：p.4
二	締約日期 (Date of signing)				參見：pp.14～16
三	締約地點 (Place of signing)				參見：p.17
四	締約當事人 (Signing parties) 及其住所或主營業所 (Address / Principal Office of signing parties)				參見：pp.17～18
五	締約當事人設立之國籍 (Nationality)				參見：pp.22～23
六	締約緣由 (Recitals / Whereas clause)				參見：pp.25～26
七	本文定義條款 (Definition)				參見：p.29
	㈠專利 (Patent)				
	㈡商標 (Trademark)				
	㈢區域 (Territory)				
	㈣授權產品 (Licensed Product)				
	㈤技術知識 (know-how)				
	㈥技術情報 (Technical Information)				
	㈦改良 (Improvement)				

	補充：				
八	本文實質規定 (Basic Conditions)				參見：p.30
	(一)技術合作內容 (Substance of Technology Transfer)				(一)(二)(三)可結合為授權條款的內容 (Grant Clause)
	(二)技術合作性質與方式 (Exclusive or Non–Exclusive, and Processes of Technology Transfer)				
	(三)授權——智慧財產權 (Grant—License of Intellectual Property Rights)				
	(四)支付報酬金 (Payment of Royalty)				
	(五)技術改良及專利的申請 (Technical Improvement and Patent Application)				
	(六)技術情報 (Technical Information)				
	(七)人員訓練 (Training of Licensee's Personnel)				
	(八)派遣技術人員 (Dispatch of Engineer)				

	㈨供應零組件 (Parts and Components Supply)				
	㈩保密條款 (Observance of Secrecy)				
	(十一)報酬 (Consideration)				
	(十二)報告 (Report)				
	(十三)侵權與救濟 (Infringement and Remedies)				
	(十四)授權人擔保／保證 (Warranties of Licensor)				
	(十五)被授權人之違約 (Default by License)				
	補充：				
九	本文一般規定 (General terms and conditions)				
	㈠契約有效期間與終止 (Duration / Period / Terms, and Termination)				參見：pp.30～31
	㈡不可抗力 (Force Majeure; Act of God)				參見：pp.32～33
	㈢轉讓 (Assignment)				參見：pp.35～38
	㈣修改 (Amendment / Modification)				參見：p.39
	㈤完整合意條款				參見：

	(Entire Agreement)				pp.40～41
	㈥翻譯／語言 (Translation / Language)				參見：pp.41～42
	㈦存續 (Survival)				參見：p.43
	㈧分離 (Severability)				參見：p.44
	㈨不棄權 (Non–Wavier)				參見：pp.44～45
	㈩通知 (Notice)				參見：pp.46～48
	⑪仲裁 (Arbitration)				參見：pp.48～49
	⑫準據法 (Applicable Law / Governing Law)與訴訟管轄 (Jurisdiction)				參見：pp.50～51
	補充：				
十	結語、簽名 (Witness clause / Signature)				參見：pp.63～65

四、【合資契約】條款檢查表

編號	檢查項目	是	否	不適用	備註
一	契約名稱 / 標題 (Title)				參見：p.5
二	締約日期 (Date of signing)				參見：pp.14～16
三	締約地點 (Place of signing)				參見：p.17
四	締約當事人 (Signing parties) 及其住所或主營業所 (Address / Principal Office of signing parties)				參見：pp.17～18
五	締約當事人設立之國籍 (Nationality)				參見：pp.22～23
六	締約緣由 (Recitals / Whereas clause)				參見：pp.25～26
七	本文定義條款 (Definition)				參見：p.29
	㈠產品 (Products)				
	㈡商標 (Trademark)				
	㈢專利 (Patents)				
	㈣區域 (Territory)				
	補充：				
	本文實質規定 (Basic Conditions)				參見：p.30

	(一)公司設立 (Organization of Company)				
	1. 公司名稱 (Name)				
	2. 公司章程				
	①本國公司章程 (Articles of Incorporation)				
	②外國公司章程 (Articles of Association)				
	3. 公司業務 (Business)				
	4. 合資目的 (Purpose)				
	5. 合資期間 (Duration)				
	6. 設立費用 (Expenses)				
	(二)公司資本 (Capital)				
八	1. 資本總額、每股價格、發行股份總數				
	2. 增資發行新股				
	(三)股份出售與轉讓之限制 (Restriction on Sale or Transfer of Shares)				
	(四)董事、監察人 (Directors and Auditors)				
	(五)股東會 (Shareholder Meeting)				
	(六)管理 (Management)				
	(七)股息、紅利 (Dividend and Bonus)				
	(八)會計 (Account Book)				

	㈨解散與清算 (Disolution and Liquidation)				
	補充：				
九	本文一般規定 (General terms and conditions)				
	㈠契約有效期間與終止 (Duration / Period / Terms, and Termination)				參見：pp.30～31
	㈡不可抗力 (Force Majeure；Act of God)				參見：pp.32～33
	㈢轉讓 (Assignment)				參見：pp.35～38
	㈣修改 (Amendment / Modification)				參見：p.39
	㈤完整合意條款 (Entire Agreement)				參見：pp.40～41
	㈥翻譯／語言 (Translation / Language)				參見：pp.41～42
	㈦存續 (Survival)				參見：p.43
	㈧分離 (Severability)				參見：p.44
	㈨不棄權 (Non–Wavier)				參見：pp.44～45
	㈩通知 (Notice)				參見：pp.46～48
	(十一)仲裁 (Arbitration)				參見：

					pp.48～49
	㈢準據法 (Applicable Law / Governing Law)與訴訟管轄 (Jurisdiction)				參見: pp.50～51
	補充:				
十	結語、簽名 (Witness clause / Signature)				參見: pp.63～65

合資公司之附屬文件可能有:

1. 【本國公司章程】－Articles of Incorporation
2. 【外國公司章程】－Articles of Association
3. 【技術合作契約書】－Technical Assistant Agreement

五、【租賃契約】條款檢查表

編號	檢查項目	是	否	不適用	備註
一	契約名稱／標題 (Title)				參見：p.5
二	締約日期 (Date of signing)				參見：pp.14～16
三	締約地點 (Place of signing)				參見：p.17
四	締約當事人 (Signing parties) 及其住所或主營業所 (Address / Principal Office of signing parties)				參見：pp.17～18
五	締約當事人設立之國籍 (Nationality)				參見：pp.22～23
六	締約緣由 (Recitals / Whereas clause)				參見：pp.25～26
七	本文定義條款 (Definition)				參見：p.29
	㈠租賃標的(The thing leased)				
	㈡				
	㈢				
	補充：				
	本文實質規定 (Basic Conditions)				參見：p.30
	㈠租金 (Rent)				

	㈡承租人之權利與義務 (The rights and Obligations of Tenant)				
	㈢出租人之權利與義務 (The rights and Obligations of Landlord)				
	㈣標的物毀損、滅失之責任 (Loss, Damages, and its Reimbursement)				
	㈤債務不履行與救濟 (Default and Remedies)				
八	㈥費用支付 (Expenses and Fees)				
	㈦擔保 (Security)				
	㈧保險 (Insurance)				
	㈨使用與轉租之限制 (Restriction on Use and Sub–let)				
	補充：				
	本文一般規定 (General terms and conditions)				
	㈠契約有效期間與終止 (Duration / Period / Terms, and Termination)				參見：pp.30～31

九	㈡不可抗力 (Force Majeure; Act of God)				參見：pp.32～33
	㈢轉讓 (Assignment)				參見：pp.35～38
	㈣修改 (Amendment/Modification)				參見：p.39
	㈤完整合意條款 (Entire Agreement)				參見：pp.40～41
	㈥翻譯／語言 (Translation / Language)				參見：pp.41～42
	㈦存續 (Survival)				參見：p.43
	㈧分離 (Severability)				參見：p.44
	㈨不棄權 (Non–Wavier)				參見：pp.44～45
	㈩通知 (Notice)				參見：pp.46～48
	(十一)仲裁 (Arbitration)				參見：pp.48～49
	(十二)準據法 (Applicable Law / Governing Law)與訴訟管轄 (Jurisdiction)				參見：pp.50～51
	補充：				
十	結語、簽名 (Witness clause / Signature)				參見：pp.63～65

六、【分期付款契約】條款檢查表

編號	檢查項目	是	否	不適用	備註
一	契約名稱／標題 (Title)				參見：p.6
二	締約日期 (Date of signing)				參見：pp.14～16
三	締約地點 (Place of signing)				參見：p.17
四	締約當事人 (Signing parties) 及其住所或主營業所 (Address / Principal Office of signing parties)				參見：pp.17～18
五	締約當事人設立之國籍 (Nationality)				參見：pp.22～23
六	締約緣由 (Recitals / Whereas clause)				參見：pp.25～26
七	本文定義條款 (Definition)				參見：p.29
	金錢單位 (Dollars)				
	補充：				
八	本文實質規定 (Basic Conditions)				參見：p.30
	㈠清償債務之方式 (Payment of Debts and Interests)				
	㈡債務視為到期				

	(Accelaration Clause)				
	㈢債務人放棄抗辯權 (Debtor's Waiver of Defense)				
	補充：				
九	本文一般規定 (General terms and conditions)				
	㈠契約有效期間與終止 (Duration / Period / Terms, and Termination)				參見： pp.30～31
	㈡不可抗力 (Force Majeure; Act of God)				參見： pp.32～33
	㈢轉讓 (Assignment)				參見： pp.35～38
	㈣修改 (Amendment / Modification)				參見：p.39
	㈤完整合意條款 (Entire Agreement)				參見： pp.40～41
	㈥翻譯／語言 (Translation / Language)				參見： pp.41～42
	㈦存續 (Survival)				參見：p.43
	㈧分離 (Severability)				參見：p.44
	㈨不棄權 (Non–Wavier)				參見： pp.44～45

	㈩通知 (Notice)				參見：pp.46～48
	⊞仲裁 (Arbitration)				參見：pp.48～49
	⊟準據法 (Applicable Law / Governing Law)與訴訟管轄 (Jurisdiction)				參見：pp.50～51
	補充：				
十	結語、簽名 (Witness clause / Signature)				參見：pp.63～65

七、【附擔保讓與契約】條款檢查表

編號	檢查項目	是	否	不適用	備註
一	契約名稱／標題 (Title)				參見：p.6
二	締約日期 (Date of signing)				參見：pp.14～16
三	締約地點 (Place of signing)				參見：p.17
四	締約當事人 (Signing parties) 及其住所或主營業所 (Address / Principal Office of signing parties)				參見：pp.17～18
五	締約當事人設立之國籍 (Nationality)				參見：pp.22～23
六	締約緣由 (Recitals / Whereas clause)				參見：pp.25～26
七	本文定義條款 (Definition)				參見：p.29
	讓與之權利 (Rights on Property Transferred)				
	補充：				
八	本文實質規定 (Basic Conditions)				參見：p.30
	㈠讓與之對價 (Consideration for Transfer)				

	(二)讓與人擔保責任 (Assignment with Warranty of Underlying Contract)				
	補充：				
九	本文一般規定 (General terms and conditions)				
	(一)契約有效期間與終止 (Duration / Period / Terms, and Termination)				參見：pp.30～31
	(二)不可抗力 (Force Majeure; Act of God)				參見：pp.32～33
	(三)轉讓 (Assignment)				參見：pp.35～38
	(四)修改 (Amendment / Modification)				參見：p.39
	(五)完整合意條款 (Entire Agreement)				參見：pp.40～41
	(六)翻譯／語言 (Translation / Language)				參見：pp.41～42
	(七)存續 (Survival)				參見：p.43
	(八)分離 (Severability)				參見：p.44
	(九)不棄權 (Non–Wavier)				參見：pp.44～45
	(十)通知 (Notice)				參見：

					pp.46～48
	(十一)仲裁 (Arbitration)				參見： pp.48～49
	(十二)準據法 (Applicable Law / Governing Law)與訴訟管轄 (Jurisdiction)				參見： pp.50～51
	補充：				
十	結語、簽名 (Witness clause / Signature)				參見： pp.63～65

八、【貸款契約】條款檢查表

編號	檢查項目	是	否	不適用	備註
一	契約名稱／標題 (Title)				參見：p.7
二	締約日期 (Date of signing)				參見：pp.14～16
三	締約地點 (Place of signing)				參見：p.17
四	締約當事人 (Signing parties) 及其住所或主營業所 (Address / Principal Office of signing parties)				參見：pp.17～18
五	締約當事人設立之國籍 (Nationality)				參見：pp.22～23
六	締約緣由 (Recitals / Whereas clause)				參見：pp.25～26
七	本文定義條款 (Definition)				參見：p.29
	金錢單位 (Dollars)				
	補充：				
八	本文實質規定 (Basic Conditions)				參見：p.30
	(一)貸款金額與用途 (Amount and Purpose)				定額、不定額借貸
	(二)貸款證明之取得				

	(Evidence of the loan as made)				
	㈢利息與償還 (Interest and Repayment)				
	㈣擔保品與處分 (Security and Disposition)				
	㈤債務不履行與救濟 (Default and Remedies)				
	㈥手續費 (Service Charges)				
	㈦保險 (Insurance)				
	補充:				
九	本文一般規定 (General terms and conditions)				
	㈠契約有效期間與終止 (Duration / Period / Terms, and Termination)				參見: pp.30～31
	㈡不可抗力 (Force Majeure; Act of God)				參見: pp.32～33
	㈢轉讓 (Assignment)				參見: pp.35～38
	㈣修改 (Amendment / Modification)				參見: p.39
	㈤完整合意條款 (Entire Agreement)				參見: pp.40～41

	㈥翻譯／語言 (Translation / Language)				參見：pp.41～42
	㈦存續 (Survival)				參見：p.43
	㈧分離 (Severability)				參見：p.44
	㈨不棄權 (Non–Wavier)				參見：pp.44～45
	㈩通知 (Notice)				參見：pp.46～48
	(十一)仲裁 (Arbitration)				參見：pp.48～49
	(十二)準據法 (Applicable Law / Governing Law)與訴訟管轄 (Jurisdiction)				參見：pp.50～51
	補充：				
十	結語、簽名 (Witness clause / Signature)				參見：pp.63～65

九、【僱傭契約】條款檢查表

編號	檢查項目	是	否	不適用	備註
一	契約名稱 / 標題 (Title)				參見：p.7
二	締約日期 (Date of signing)				參見：pp.14～16
三	締約地點 (Place of signing)				參見：p.17
四	締約當事人 (Signing parties) 及其住所或主營業所 (Address / Principal Office of signing parties)				參見：pp.17～18
五	締約當事人設立之國籍 (Nationality)				參見：pp.22～23
六	締約緣由 (Recitals / Whereas clause)				參見：pp.25～26
七	本文定義條款 (Definition)				參見：p.29
	補充：				
八	本文實質規定 (Basic Conditions)				參見：p.30
	㈠受僱場所 (Place)				
	㈡工作時間與休假 (Work Time and Vocation)				
	㈢報酬 (Compensation)				

	㈣工作費用之補償 (Reimbursement for Expense)				
	㈤保密 (Confidentiality)				
	㈥僱用人之義務與責任 (Employer Responsibilities and Duties)				
	㈦受僱人之義務與責任 (Employee Responsibilities and Duties)				
	㈧員工福利 (Benefits)				
	㈨違約與救濟 (Breach and Remedies)				
	補充：				
九	本文一般規定 (General terms and conditions)				
	㈠契約有效期間與終止 (Duration / Period / Terms, and Termination)				參見： pp.30～31
	㈡不可抗力 (Force Majeure; Act of God)				參見： pp.32～33
	㈢轉讓 (Assignment)				參見： pp.35～38
	㈣修改 (Amendment /				參見：p.39

	Modification)				
	㈤完整合意條款 (Entire Agreement)				參見： pp.40～41
	㈥翻譯／語言 (Translation / Language)				參見： pp.41～42
	㈦存續 (Survival)				參見：p.43
	㈧分離 (Severability)				參見：p.44
	㈨不棄權 (Non–Wavier)				參見： pp.44～45
	㈩通知 (Notice)				參見： pp.46～48
	⑪仲裁 (Arbitration)				參見： pp.48～49
	⑫準據法 (Applicable Law / Governing Law)與訴訟管轄 (Jurisdiction)				參見： pp.50～51
	補充：				
十	結語、簽名 (Witness clause / Signature)				參見： pp.63～65

十、【保證契約】條款檢查表

編號	檢查項目	是	否	不適用	備註
一	契約名稱／標題 (Title)				參見：p.7
二	締約日期 (Date of signing)				參見：pp.14～16
三	締約地點 (Place of signing)				參見：p.17
四	締約當事人 (Signing parties) 及其住所或主營業所 (Address / Principal Office of signing parties)				參見：pp.17～18
五	締約當事人設立之國籍 (Nationality)				參見：pp.22～23
六	締約緣由 (Recitals / Whereas clause)				參見：pp.25～26
七	本文定義條款 (Definition)				參見：p.29
	補充：				
	本文實質規定 (Basic Conditions)				參見：p.30
	㈠保證性質 (The Nature)				例如：保證係屬絕對的 (Absolute)；不可撤銷的

八					(Irrevocable)；無條件的(Uncondi-tinoal)；繼續性的(Continuing)
八	(二)保證效力 (Effect)				
八	(三)付款擔保 (Security for payment)				
八	(四)責任限制 (Limitation on liability)				
八	(五)代位清償 (Subrogation)				
八	(六)債務證據 (Evidence of Debt)				
八	補充：				
九	本文一般規定 (General terms and conditions)				
九	(一)契約有效期間與終止 (Duration / Period / Terms, and Termination)				參見：pp.30～31
九	(二)不可抗力 (Force Majeure；Act of God)				參見：pp.32～33
九	(三)轉讓 (Assignment)				參見：pp.35～38

	㈣修改 (Amendment / Modification)				參見：p.39
	㈤完整合意條款 (Entire Agreement)				參見：pp.40～41
	㈥翻譯／語言 (Translation / Language)				參見：pp.41～42
	㈦存續 (Survival)				參見：p.43
	㈧分離 (Severability)				參見：p.44
	㈨不棄權 (Non–Wavier)				參見：pp.44～45
	㈩通知 (Notice)				參見：pp.46～48
	�(十一)仲裁 (Arbitration)				參見：pp.48～49
	�(十二)準據法 (Applicable Law / Governing Law)與訴訟管轄 (Jurisdiction)				參見：pp.50～51
	補充：				
十	結語、簽名 (Witness clause / Signature)				參見：pp.63～65

十一、【不動產抵押契約】條款檢查表

編號	檢查項目	是	否	不適用	備註
一	契約名稱／標題 (Title)				參見：p.8
二	締約日期 (Date of signing)				參見：pp.14～16
三	締約地點 (Place of signing)				參見：p.17
四	締約當事人 (Signing parties) 及其住所或主營業所 (Address / Principal Office of signing parties)				參見：pp.17～18
五	締約當事人設立之國籍 (Nationality)				參見：pp.22～23
六	締約緣由 (Recitals / Whereas clause)				參見：pp.25～26
七	本文定義條款 (Definition)				參見：p.29
	㈠抵押標的 (The property subject to mortgage)				
	㈡金錢單位 (Dollars)				
	補充：				
	本文實質規定 (Basic Conditions)				參見：p.30
	㈠抵押條件 (The Terms)				

八	㈡抵押金額 (The Amount)				
	㈢債務返還 (Repayment)				
	㈣債務不履行與救濟 (Default and Remedies)				
	㈤抵押物孳息 (Interest or other return)				
	㈥保管抵押物 (Care of mortgaged property)				
	㈦處分抵押物 (Disposition)				
	㈧更換抵押物 (Subsitution in Collateral)				
	㈨保險 (Insurance)				
	㈩拍賣抵押物 (Sale)				
	補充：				
九	本文一般規定 (General terms and conditions)				
	㈠契約有效期間與終止 (Duration / Period / Terms, and Termination)				參見：pp.30～31
	㈡不可抗力 (Force Majeure；Act of God)				參見：pp.32～33
	㈢轉讓 (Assignment)				參見：pp.35～38
	㈣修改 (Amendment /				參見：p.39

	Modification)				
	㈤完整合意條款 (Entire Agreement)				參見：pp.40～41
	㈥翻譯／語言 (Translation / Language)				參見：pp.41～42
	㈦存續 (Survival)				參見：p.43
	㈧分離 (Severability)				參見：p.44
	㈨不棄權 (Non–Wavier)				參見：pp.44～45
	㈩通知 (Notice)				參見：pp.46～48
	∪仲裁 (Arbitration)				參見：pp.48～49
	∫準據法 (Applicable Law / Governing Law)與訴訟管轄 (Jurisdiction)				參見：pp.50～51
	補充：				
十	結語、簽名 (Witness clause / Signature)				參見：pp.63～65

十二、【營業秘密契約】條款檢查表

編號	檢查項目	是	否	不適用	備註
一	契約名稱／標題 (Title)				參見：p.9
二	締約日期 (Date of signing)				參見：pp.14～16
三	締約地點 (Place of signing)				參見：p.17
四	締約當事人 (Signing parties) 及其住所或主營業所 (Address / Principal Office of signing parties)				參見：pp.17～18
五	締約當事人設立之國籍 (Nationality)				參見：pp.22～23
六	締約緣由 (Recitals / Whereas clause)				參見：pp.25～26
七	本文定義條款 (Definition)				參見：p.29
	營業秘密 (Trade Secrect)				
	補充：				
八	本文實質規定 (Basic Conditions)				參見：p.30
	㈠保密事項 (Terms and Conditions)				
	㈡保密義務				

	(Non–Discloure Duties)				
	㈢違反與救濟 (Breach and Remedy)				
	補充：				
九	本文一般規定 (General terms and conditions)				
	㈠契約有效期間與終止 (Duration / Period / Terms, and Termination)				參見： pp.30～31
	㈡不可抗力 (Force Majeure; Act of God)				參見： pp.32～33
	㈢轉讓 (Assignment)				參見： pp.35～38
	㈣修改 (Amendment / Modification)				參見：p.39
	㈤完整合意條款 (Entire Agreement)				參見： pp.40～41
	㈥翻譯／語言 (Translation / Language)				參見： pp.41～42
	㈦存續 (Survival)				參見：p.43
	㈧分離 (Severability)				參見：p.44
	㈨不棄權 (Non–Wavier)				參見： pp.44～45
	㈩通知 (Notice)				參見：

					pp.46～48
	(十一)仲裁 (Arbitration)				參見： pp.48～49
	(十二)準據法 (Applicable Law / Governing Law)與訴訟管轄 (Jurisdiction)				參見： pp.50～51
	補充：				
十	結語、簽名 (Witness clause / Signature)				參見： pp.63～65

十三、【信託契約】條款檢查表

編號	檢查項目	是	否	不適用	備註
一	契約名稱／標題 (Title)				參見：p.9
二	締約日期 (Date of signing)				參見：pp.14～16
三	締約地點 (Place of signing)				參見：p.17
四	締約當事人 (Signing parties) 及其住所或主營業所 (Address / Principal Office of signing parties)				參見：pp.17～18
五	締約當事人設立之國籍 (Nationality)				參見：pp.22～23
六	締約緣由 (Recitals / Whereas clause)				參見：pp.25～26
七	本文定義條款 (Definition)				參見：p.29
	㈠委託人 (Grantor)				
	㈡受託人 (Trustee)				
	㈢受益人 (Beneficiaries)				
	補充：				
	本文實質規定 (Basic Conditions)				參見：p.30
	㈠信託資產 (The Trust				

	Estate)				
	㈡撤回；更改 (Revocation; Modification)				
	㈢撤回或更改權 (Power to Revoke or Modification)				
	㈣對受託人之特別撤回權 (Special Power of Revocation to Trustee)				
	㈤增減受益人之權 (Power to Add / Exclude Beneficiaries)				
八	㈥受益人與財產分配 (Beneficiaries and Distribution)				
	㈦委託人生存時之信託 (The Trusts during Grentor's Lifetime)				
	㈧委託人死後之信託 (Trusts after Grantor's Death)				
	㈨受託人權限 (Power of Trustee)				
	補充：				
	本文一般規定 (General terms and				

	conditions)				
	㈠契約有效期間與終止 (Duration / Period / Terms, and Termination)				參見：pp.30～31
	㈡不可抗力 (Force Majeure; Act of God)				參見：pp.32～33
	㈢轉讓 (Assignment)				參見：pp.35～38
	㈣修改 (Amendment / Modification)				參見：p.39
	㈤完整合意條款 (Entire Agreement)				參見：pp.40～41
	㈥翻譯／語言 (Translation / Language)				參見：pp.41～42
九	㈦存續 (Survival)				參見：p.43
	㈧分離 (Severability)				參見：p.44
	㈨不棄權 (Non–Wavier)				參見：pp.44～45
	㈩通知 (Notice)				參見：pp.46～48
	⑴仲裁 (Arbitration)				參見：pp.48～49
	⑵準據法 (Applicable Law / Governing Law)與訴訟管轄 (Jurisdiction)				參見：pp.50～51
	補充：				

十	結語、簽名 (Witness clause / Signature)				參見: pp.63～65

注意

信託契約 (Fiduciary Contract; Trust Agreement)
信託契據 (Trust Deeds)
信託收據 (Trust Receipt)
信託書 (Letter of Trust)
信託證書 (Trust Certificate)
信託聲明 (Declaration of Trust)

◈附錄二

契約格式中英對照一覽表

契　約

本契約由A公司（以下簡稱A）——總公司位於住址，依照某國之法律設立；與B公司（以下簡稱B）——總公司位於住址，依照某國之法律設立，於　年　月　日，地點，訂定契約。

茲証明

鑒於＿＿＿＿＿＿＿＿＿＿

與

鑒於＿＿＿＿＿＿＿＿＿＿

因此，以＿＿＿為約因，同意如下：

第一條

第二條

第三條

⋮

本契約一式二份，由雙方正式授權其承辦人員或代表，於首開日期訂定生效。

甲方：×××公司	乙方：×××公司
＿＿＿＿＿＿＿＿	＿＿＿＿＿＿＿＿
簽名	簽名
＿＿＿＿＿＿＿＿	＿＿＿＿＿＿＿＿
姓名：×××	姓名：×××
＿＿＿＿＿＿＿＿	＿＿＿＿＿＿＿＿
職稱	職稱
＿＿＿＿＿＿＿＿	＿＿＿＿＿＿＿＿
日期	日期
封印	封印

AGREEMENT

THIS AGREEMENT, (is) made and entered into in 地點 this 日期（序數）day of 月, 年 by and between A (hereinafter called “A”), a corporation duly organized and existing under the laws of 某國, having its principal office at 住址, and B (hereinafter called “B”), a corporation duly organized and existing under the laws of 某國, having its principal office at 住址.

WITNESSETH

WHEREAS,..,

and

WHEREAS, ...,

NOW THEREFORE, in consideration of ...., it is agreed as follows:

Article 1

⋮

IN WITNESS WHEREOF, this agreement has been executed by the duly authorized officers or representatives of the parties in duplicate as of the date first above written.

Party A： ____________	Party B： ____________
____________	____________
Signature	Signature
____________	____________
Printed Name	Printed Name
____________	____________
Title	Title
____________	____________
Date	Date
L.S. / Seal	L.S. / Seal

末了的話

法律能反應出不同國籍、種族與文化的人，在不同的時空背景下，不同的價值觀。而商業活動的盛衰不難由相關商業法規中窺知一二。在商業活動頻繁的現代，契約種類何其眾多，然而，事實上，我國民法所制訂商業上契約的種類仍然無法完全包含並滿足一切商業活動的需求。因此，當事人間在商業活動上所應遵循的權利義務，仍有賴於不同具體案例建置並落實於契約條款內容中。

現代聰明的企業經營者莫不將「成本與風險降低，以追求最高利潤」奉為圭臬。相信在商務上，有部分企業經營者已察覺到謹慎、嚴謹地面對商務契約，能避免許多不必要的紛爭與時間、勞力、費用的損失及浪費。而面對非國人母語的英文商務契約，這種對契約的謹慎、嚴謹態度，更應被格外地要求與強調！

最後，希望這本書對讀者有所貢獻，對本書如有任何疑問與指正，敬請各位讀者與先進不吝賜教。歡迎來電、來信或上網與我們聯繫。我們的聯絡方式如下：

惠聯地政商務法律事務所（Jusleader Law Offices）

【台北所】：台北市羅斯福路三段300號6樓

Tel:（02）2369-8979

Fax:（02）2368-0889

Taipei Office: 6F, No. 300, Sec. 3, Roosevelt Rd., Taipei, Taiwan

【士林所】：台北市忠誠路一段119號2樓

Tel:（02）2831-1289

Fax:（02）2832-2451

Shihlin Office: 2F, No. 119, Sec. 1, Jungcheng Rd., Taipei, Taiwan

【桃園所】：桃園縣中壢市中正路507號2樓

Tel:（03）492-9308

Fax:（03）494-5110

Taoyuan Office: 2F, No. 507,Jungjeng Rd., Jungli City, Taoyuan Hsien, Taiwan

http://www.wake-landgroup.com.tw

E-mail: wakeland@ms47.hinet.net

主要參考書籍

張錦源，民國84年，貿易契約理論與實務，第四版，三民書局。

陳春山，民國87年，國際商務契約，再版，三民書局。

朱鳳仙譯，民國86年，契約英語，初版，台灣英語雜誌社。

五南編輯部，民國84年，法律英漢辭典，初版五刷，五南圖書出版有限公司。

薛波主編，民國90年，法律漢英辭典，初版，五南圖書出版有限公司。

Steven H. Gifis, 1984, Law Dictionary, Barron's Educational Series, Inc.

◎ 貿易英文撰寫實務　張錦源／著

本書係作者以其多年從事外匯、貿易的經驗撰寫而成，全書首先介紹貿易英文信函的結構與貿易英文文法，其次循進出口貿易的程序，將進出口商在每一階段的往來函電，舉實例說明其撰寫要領及應注意事項，並從貿易實務觀點作詳盡的注釋。透過本書，讀者不僅能瞭解貿易英文函電的寫作要領，亦可學到貿易實務的技巧。

◎ 英文貿易契約撰寫實務　張錦源／著

本書作者參考中外名著及教學心得，從法律觀點闡明貿易契約之意義及重要性、貿易契約條款之結構及各種契約用語，以及各種貿易慣例。實務方面則說明如何撰寫貿易契約書、經銷契約書、國外合資契約書等。如能仔細閱讀，可訂立各種完善之貿易契約書，防範無謂之貿易糾紛，開展貿易業務。

◎ 國貿實務全國會考教材

國貿大會考教材編審委員會／編著

本書係依據「國際貿易大會考」測驗內容編輯而成，分別介紹國際貿易基本概念、進出口流程、貿易條件、交易基本條件、交易磋商與契約成立、國際貿易付款方式、國際貨物運輸、國際貿易風險管理、進出口結匯與提貨、進出口價格計算與貿易單據製審及我國外貿現況與發展。

◎ 國貿業務丙級檢定學術科試題解析

康蕙芬／編著

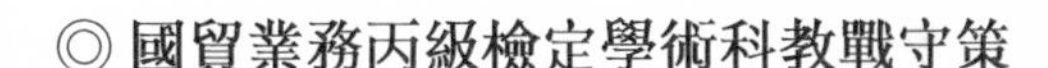

本書係依據勞委會公告之國貿業務丙級技能檢定學科題庫與術科範例題目撰寫，學科部分將題庫 700 題選擇題，依據貿易流程的先後順序先作重點整理、分析，再就較難理解的題目進行解析。術科部分依據勞委會公告之範例，以五個章節分別解說；首先提示重點與說明解題技巧，接著附上範例與解析，最後並有自我評量單元供讀者練習。

◎ 國貿業務丙級檢定學術科教戰守策

張　瑋／編著

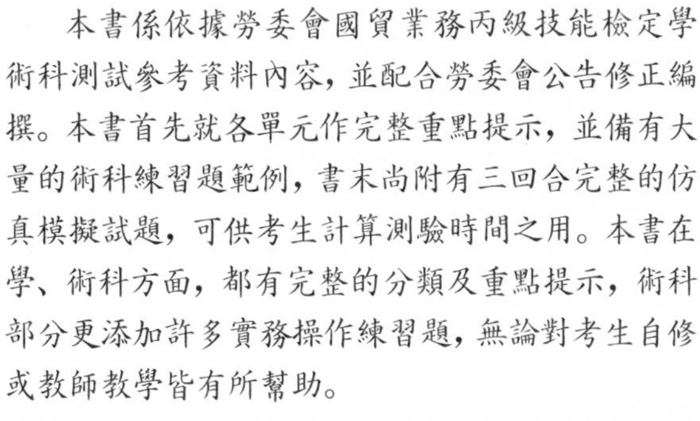

本書係依據勞委會國貿業務丙級技能檢定學術科測試參考資料內容，並配合勞委會公告修正編撰。本書首先就各單元作完整重點提示，並備有大量的術科練習題範例，書末尚附有三回合完整的仿真模擬試題，可供考生計算測驗時間之用。本書在學、術科方面，都有完整的分類及重點提示，術科部分更添加許多實務操作練習題，無論對考生自修或教師教學皆有所幫助。

◎ 國際貿易實務　　張錦源、劉　玲／編著

本書以簡明淺顯的筆法闡明國際貿易的進行程序，並附有周全的貿易單據，如報價單、輸出入許可證申請書、郵遞信用狀、電傳信用狀、商品輸出檢驗申請書、海運提單、空運提單、領事發票及保結書等，同時有填寫方式與注意事項等說明，再輔以實例連結，增加讀者實務運用的能力；本書於每章之後，均附有豐富的習題，以供讀者評量閱讀本書的效果。

◎ 國際企業管理　　陳弘信／著

國際企業經營管理涉及層面廣且深，有鑑於此，本書綜合各領域，歸納成國際經濟與環境、國際金融市場、國際經營與策略、國際營運管理四大範疇說明。在內容編排上，每章都附有架構圖，並列有學習重點，條列探討主題。另外配合實務個案編有引導教學，讀者可先有概括性認識，再配合關鍵思考的提醒，有利於後續章節內容的瞭解。在章末則安排個案問題與討論，讓讀者運用所學，進行邏輯思考與應用。

◎ 國際貿易理論與政策

歐陽勛、黃仁德／著

本書乃為因應研習複雜、抽象之國際貿易理論與政策而編寫。對於各種貿易理論的源流與演變，均予以有系統的介紹、導引與比較，採用大量的圖解，作深入淺出的剖析，由靜態均衡到動態成長，由實證的貿易理論到規範的貿易政策，均有詳盡的介紹。

◎ 經濟學　　王銘正／著

作者大量利用實務印證與鮮活例子，使讀者能充分領略本書所介紹的內容。在全球金融整合程度日益升高之際，國際金融知識也愈顯重要，因此本書運用相當的篇幅介紹「國際金融」知識，並利用相關理論說明臺灣與日本的「泡沫經濟」，以及「亞洲金融風暴」。本書也在每一章的開頭列舉該章的學習重點，方便讀者對每一章的內容建立起基本概念，也提供讀者在複習時自我檢視學習成果。

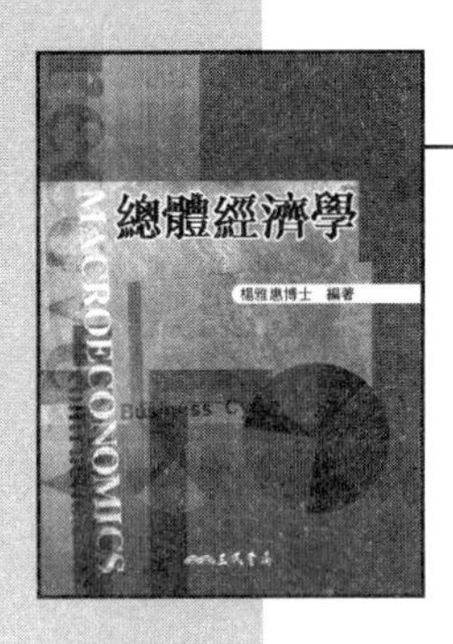

◎ 總體經濟學　　楊雅惠／編著

總體經濟學是用來分析總體經濟的知識與工具，而如何利用其基本架構，來剖析經濟脈動、研判經濟本質，乃是一大課題。一般總體經濟學書籍缺乏實務分析或是案例，本書即著眼於此，特別在每章內文中巧妙導入臺灣之經濟實務資訊，如民生痛苦指數、國民所得統計等相關實際數據。在閱讀理論部分後，讀者可以馬上利用實際數據與實務接軌，這部分將成為讀者在日後進行經濟分析之學習基石。

◎ 貨幣銀行學　　楊雅惠／編著

本書介紹貨幣銀行學，內容涵蓋貨幣概論、金融體系、銀行業與金融發展、貨幣供給、貨幣需求、利率理論、總體貨幣理論、央行貨幣政策，與國際金融等篇。

每章均採用架構圖與有層次的標題來引導讀者建立整體的概念，並配合各章節理論之介紹，引用臺灣最新近的金融資訊來佐證，期能讓理論與實際之間互相結合，因此相當適合初學者入門，再學者複習，實務者活用。